A PLU

HOW TO BUI

JACK HORNER is a regents' professor of paleontology at Montana State University, and probably the best-known paleontologist in the world. He is the recipient of a MacArthur "Genius" Award and is the author of several books on dinosaurs. He lives in Bozeman, Montana.

JAMES GORMAN is deputy science editor of *The New York Times*. He lives near New York City.

"*How to Build a Dinosaur* . . . indicates how paleontology is becoming more of an interdisciplinary science, where studies of development and genetics are just as important as fossilized bones. The bodies of living things hold records of the past just as the strata of the earth do, and when both lines of evidence are studied together scientists can finally begin to answer evolutionary questions that have puzzled researchers for decades." —*Smithsonian*

"Brash, world-renowned paleontologist [Horner] confesses that if he had his druthers, he'd 'Bring 'em Back Alive.' . . . A few centuries back, he would have been burned at the stake for this suggestion; today, it's exciting. Evo devo for the everyday reader, with the personal stuff adding color needed to sustain a skirmish with molecular paleobiology." —*Kirkus Reviews*

"The authors raise questions about the ethics of altering embryos and debunk the notion that a Jurassic Park could ever really be created. But they also don't rule out the possibility of giving a chicken a few dinosaur features, as one McGill University scientist is attempting to do. Sure to appeal to dinosaur fans, this readable account of innovative science is recommended for public as well as academic library collections." —*Library Journal*

HOW TO BUILD A
DINOSAUR

THE NEW SCIENCE OF REVERSE EVOLUTION

JACK HORNER
and JAMES GORMAN

A PLUME BOOK

PLUME
Published by the Penguin Group
Penguin Group (USA) Inc., 375 Hudson Street, New York, New York 10014, U.S.A. • Penguin
Group (Canada), 90 Eglinton Avenue East, Suite 700, Toronto, Ontario, Canada M4P 2Y3 (a divi-
sion of Pearson Penguin Canada Inc.) • Penguin Books Ltd., 80 Strand, London WC2R 0RL,
England • Penguin Ireland, 25 St. Stephen's Green, Dublin 2, Ireland (a division of Penguin
Books Ltd.) • Penguin Group (Australia), 250 Camberwell Road, Camberwell, Victoria 3124,
Australia (a division of Pearson Australia Group Pty. Ltd.) • Penguin Books India Pvt. Ltd., 11
Community Centre, Panchsheel Park, New Delhi – 110 017, India • Penguin Group (NZ), 67
Apollo Drive, Rosedale, North Shore 0632, New Zealand (a division of Pearson New Zealand
Ltd.) • Penguin Books (South Africa) (Pty.) Ltd., 24 Sturdee Avenue, Rosebank, Johannesburg
2196, South Africa

Penguin Books Ltd., Registered Offices: 80 Strand, London WC2R 0RL, England

Published by Plume, a member of Penguin Group (USA) Inc. Previously published in a
Dutton edition.

First Plume Printing, March 2010

10   9   8   7   6   5   4   3   2   1

Photo credits: Page 30, courtesy of the author; Page 82, Mary Schweitzer; Page 195,
© J. J. Audubon/VIREO; Pages 216 and 217, © Phil Wilson.

Ⓟ REGISTERED TRADEMARK—MARCA REGISTRADA

The Library of Congress has catalogued the Dutton edition as follows:

Horner, John R.
    How to build a dinosaur : extinction doesn't have to be forever /
Jack Horner and James Gorman.
       p.   cm.
    Includes index.
    ISBN 978-0-525-95104-9 (hc.)
    ISBN 978-0-452-29601-5 (pbk.)
    1. Evolutionary paleobiology.   2. Dinosaurs—Extinction.   I. Gorman, James,
1949–   II. Title.
    QE721.2.E85H67 2009
    567.9—dc22   2008048042

Printed in the United States of America
Original hardcover design by Daniel Lagin
Set in Dante MT

For Darwin

CONTENTS

HOW TO BUILD A DINOSAUR

INTRODUCTION

Nothing is too wonderful to be true if it be consistent with the laws of nature, and in such things as these, experiment is the best test of such consistency.

—Michael Faraday

Let's suppose you wanted to pick a moment in the history of life and play it over again, backward and forward, like a football play on a highlights DVD, so you could see exactly how it happened. Rewind. Stop. Play. Rewind frame by frame. Stop. Play frame by frame.

Stephen Jay Gould, one of the best-known evolutionary biologists of his time, wrote in *Wonderful Life*, his book on the weird and wonderful fossils of a rock formation known as the Burgess Shale, that you can't go home again, evolutionarily, unless you want to risk not being here when you come back. What he was saying was that evolution is a chance business, contingent on many influences and events. You can't rewind it and run it over and hope to get the same result. The second time through *Homo sapiens* might not appear. Primates might not appear.

That's evolution on a grand scale, major trends in the history of life that involved mass extinctions and numerous species jockeying for evolutionary position. We can't rewind that tape without a planet to toy with. But I'm thinking about a time machine with a somewhat closer focus, an evolutionary microscope that could target, say, the first appearance of feathers on dinosaurs, or the evolution of dinosaurs into birds.

This time machine/microscope could zero in on one body part. For birds we might start small, with a much maligned body part—the tail. We don't think about tails much, not at the high levels of modern evolutionary biology, but they are more intriguing than you might imagine. They appear and disappear in evolution. They appear and disappear in the growth of a tadpole. Most primates have tails. Humans and great apes are exceptions.

The dinosaurs had tails, some quite remarkable. Birds, the descendants of dinosaurs, now almost universally described by scientists as avian dinosaurs, do not have tails. They have tail feathers but not an extended muscular tail complete with vertebrae and nerves. Some of the first birds had long tails, and some later birds had short tails. But there is no modern bird with a tail.

How did that change occur? Is there a way to re-create that evolutionary change and see how it happened, right down to the molecules involved in directing, or stopping, tail growth?

I think the answer is yes. I think we can rewind the tape of bird evolution to the point before feathers or a tail emerged, or teeth disappeared. Then we can watch it run forward, and then rewind again, and try to play it without the evolutionary change, reverting to the original process. I'm not suggesting we can do this on a grand scale, but we can pick a species, study its growth as an embryo, learn how it develops, and learn how to change that development.

Then we can experiment with individual embryos, intervening in development in different ways—with no change, with one change, or several changes. This would be a bit like redoing Game 6 of the 1986 World Series between the New York Mets and the Boston Red Sox, when a ground ball ran between first baseman Bill Buckner's legs and changed the tide of the series.

We would be doing more than just fiddling with the tape; we would be redoing the play, with Bill Buckner and all the players. And the idea would be to determine the precise cause of the Mets' joy and Red Sox' sadness. Was it Buckner's failing legs, the speed of the ball, the topography of the field? What caused him to miss the ball? And when we think we know the cause, we test our hypothesis. We give him younger legs or smooth out the field and then we see if in this altered set of circumstances, he snags the grounder.

That's impossible to do in baseball. We don't have a way to go back in time. We can do it with computer models, of course, in both baseball and biology. But with current technology and our current understanding of development and evolution, we could also do it with a living organism. This ability is largely the result of a new and thriving field of research that has joined together the study of how an embryo develops with the study of how evolution occurs. The idea, in simple terms, is that because the shape or form of an animal emerges as it grows from a fertilized egg to hatching or birth, any evolutionary change in that shape must be reflected in a change in the way the embryo grows.

For example, in a long-tailed ancient bird embryo the tail would have started to develop and continued to develop until the chick hatched with a full tail. The embryos of descendant species, which hatched with no tails, would have to develop in a different way. We can observe the embryos of modern birds as they develop, and if we can pinpoint the moment at which the tail stops growing, we can figure out exactly what events occurred at the molecular level to stop tail growth. We can say—that's where the change occurred in evolution. And it is an idea we can test. We can try intervening at that moment in the embryo's growth to change the growth and development signals back to what we believe they were before the tail disappeared in evolution. If we are right, then the long tail should grow. If we can do this with a tail, we ought to be able to do it with teeth, feathers, wings, and feet.

The most studied and most available bird for both laboratory and culinary experiments is the chicken. Why couldn't we take a chicken embryo and biochemically nudge it this way and that, until what hatched was not a chicken but a small dinosaur, with teeth, forearms with claws, and a tail? No reason at all.

We haven't done it yet. But we are taking the first small

steps. This book is about those steps, the path ahead, what we could learn, and why we should do this experiment.

Hatching a dinosaur from a chicken's egg may sound like something that belongs in a movie. It seems very remote from my specialty, vertebrate paleontology, in particular the study of dinosaurs. Paleontologists, after all, are the slightly eccentric folks who dig up old bones in sun-drenched badlands and like to talk too much about skulls and femurs. Well, that may be true, as far as it goes. But that's only part of the story. At heart, every paleontologist is as much Frank Buck as Stephen Jay Gould.

Frank Buck was a real person who became a hero of movies and books before and after World War II. He went into the jungles and remote places of the world and brought back not fossils, but exotic living animals. He was—and this appealed to many a small boy—not a hunter who killed his prey, but a collector of live animals. And his motto, once as well known as any of today's catchphrases, was: Bring 'em back alive.

Well, paleontologists may deal with the long dead. But at the heart of all the digging and preparation of skeletons and museum displays is the attempt to reconstruct the past, to re-create moments in the history of life. What we would really love to do, if we could, is bring 'em back alive.

That hasn't really been possible for the past two centuries as dinosaur scientists have ventured far into the past, into what is often called deep time, millions and millions of years ago, retrieving clues and broken pieces of a puzzle that we then try to solve.

Sometimes it's a literal puzzle, with pieces of fossil skeletons that must be connected and made to fit. Sometimes it's a puzzle at another level, trying to put together a long-gone ecological system. Sometimes it is macroscopic—writing the story of the great trends of evolution, from the seas to land, from the land to the air, from reptile to bird. Sometimes it is microscopic, digging deep into the tissue of ancient bones to tease out the physiology of dinosaurs, or the molecular makeup of fossilized tissue. We describe what dinosaurs were like and present these ideas in scientific papers and books. We build skeletons and sculptures of dinosaurs that any museumgoer can appreciate. We have made robotic dinosaurs for education and entertainment. We have even helped make movies hew more closely to the scientific facts. So it's a natural enough step to go from building a dinosaur to growing one, from the mechanical to the biological. Or so it seemed, and seems to me, which is how I became an instigator and a recruiter, looking for scientists with more expertise than me in the molecular biology lab to pursue what may seem at first like a cockeyed idea, to make the ultimate reconstruction of the past, a living dinosaur.

The first part of any reconstruction is to understand just what it is you are trying to make. If you're going to indulge in biological reverse engineering, you have to take your target creature apart to see how it works. That's the bread and butter of dinosaur scientists. We find fossils, dig them up, date them, put them in a context with other organisms that lived during their time. We use the fossilized bones to establish the shape of the dinosaur. We make educated guesses, some more solid than others, about movement, behavior, social life, parental involvement with the young.

With modern imaging technologies and computing power, we look deeper than ever before into the fossils we find. We can see inside the bones. We use CT scans to make 3-D images of the inside of skulls. We smash up bits of fossils to search for preserved remnants of muscle tissue, blood vessels, red blood cells. We use the tools of chemistry and physics to go deeper yet.

The recent technological changes in how bones are studied are profound. For most of the last century the study of dinosaurs was primarily a collector's game. It certainly was not an experimental science. But that is changing. We can now retrieve ancient biomolecules, like proteins, from fossils tens of millions of years old. And we can mine the genomes of living creatures to trace evolutionary history. We can bring the history of life into the laboratory to test our ideas with experiments. And right at the top of the list of the experiments we can try is the attempt to bring back the characteristics of extinct creatures that have long been lost to us in deep time. That is how we can build a dinosaur.

We may someday recover bits of dinosaur DNA, but that is not the route to making a dinosaur. That has already been tried, in the movies. But it won't happen in real life. I'm not putting down the movies. I loved *Jurassic Park,* not least because I worked on it and the sequels as a technical consultant to help get the dinosaurs right. And the idea of cloning a dinosaur from DNA recovered from a mosquito preserved in amber that once fed on dinosaurs was a brilliant fiction. It was, however, a fiction that reflected the science of its time, the fascination with DNA and the idea that we would have a complete blueprint of a dinosaur to make one. Now we are actually much closer to being able to create a dinosaur, without needing to recover ancient DNA.

We can do it because of the nature of evolution, and the way it builds on itself, adapting old plans to new circumstances, not inventing new life-forms from scratch. Much of the writing about dinosaurs in recent times has concentrated on the way we have been correcting our old mistakes. But mistakes are to be expected when you are trying to reach back tens of millions of years. What is amazing, if you stop to think about it, is that we got the main points right about their shape and structure right off the bat. How could we do that so easily when you might think that such ancient animals could have taken any shape imaginable, or unimaginable?

The answer is that evolution does not allow innovation without limit. It did not allow the dinosaurs to pop up in any old shape. They have the same body plan that all other animals with backbones do. Anyone can see that immediately, with or without science. Anyone who has seen a deer skeleton, or a lizard skeleton, or a human skeleton, would recognize the basics in a bunch of *T. rex* bones dug from the ground. *T. rex* has a spine, a skull, ribs, just as an eel or a salmon or a mouse does. It has hind limbs and forelimbs, just as crocodiles and frogs, hawks and people do.

Why? Why do animals of such different external shapes and lives share characteristics so similar that we can immediately recognize the basics of their bone structure? The reason is that shape is not an unlimited smorgasbord from which evolution can pick and choose. All living things are part of a continuum. The shapes of animals evolve over time from earlier ancestral shapes. Different groups of animals have basic body plans that themselves evolved from earlier plans.

All vertebrates have backbones. But before that innovation, they evolved from creatures that all had a front-to-back orientation for eating and elimination. Before that came self-propulsion. Before that energy metabolism. And so on, back to DNA itself, which we share with all living things, unless RNA viruses count as living things. A body plan is an abstract idea—four limbs, spine, skull, mouth at the front, elimination at the other end—but it has remained constant and resilient. Mountains have risen and fallen, seas have appeared and dried up, and continents themselves have shifted, while the standard four-limb plan, the tetrapod blueprint, has persisted with minor modifications.

The bones in the hands and arms used to type these words are almost the same as the bones in Buffalo chicken wings. If you follow the development of a chicken embryo closely, you will see five buds at the end of the developing wing, buds that also appear in the embryos of mice and people. The buds become fingers in a human embryo and claws in a mouse. In a chicken the five buds on a forelimb will lengthen, shorten, disappear, and be fused to fit into the familiar structure that cries out to us for hot sauce. The record of such astonishing and persistent continuity of form has been described as the main gift the fossil record has given to evolutionary biology.

Evolutionary change is added to existing plans. Genetic blueprints are not thrown out. We don't have to start from scratch to grow a dinosaur. We don't have to retrieve ancient DNA for cloning. Birds are descended from dinosaurs. Actually, they *are* dinosaurs, and most of the genetic program for the dinosaur characteristics we want to bring back should still be available in birds—in fact, in the chicken.

You can see evidence of this continuity in the way an embryo develops. And chicken embryos, in those perfectly functional containers, hard-shelled eggs, have been endlessly studied. Aristotle was the first to record the stages of growth of a chicken embryo. Other scientists have followed his example, partly because chickens and chicken eggs are so readily available.

You can see easily and clearly with a low-power dissecting microscope that a tail like a dinosaur's is well on its way in the growing chicken embryo before something stops it. The result is a plump tail stump called the pope's or parson's nose (if your specialty is eating chickens), or the pygostyle (if you are more given to studying them). The pygostyle is a hodgepodge of different bones, the growth and purpose of which have been redirected, just the kind of jumble that evolution specializes in. It is a perfect demonstration that evolution does not suggest intelligence, planning, or purpose, but rather accident and opportunism. Evolution is by definition not revolution. It works within the system, using what it finds.

One of the hottest fields in science now is evo-devo, for "evolutionary developmental biology." It has also been called devo-evo, or DE, for "developmental evolution." Whatever the name, it is the investigation of how evolution proceeds through changes in the growth of embryos. The limb-to-wing transition does not require a complete new set of genes, but rather changes in the control of a few genes that promote or stop growth. These genes produce chemicals called growth and signaling factors that give directions to the cells in a growing embryo. When they are turned on and off at different times, that can drastically change the shape of an animal.

It's a bit like remixing an old recording. Let's take a band with a banjo, guitar, and a mandolin playing "She'll Be Coming 'Round the Mountain." The original recording features the banjo, but you want just the guitar and mandolin, so you turn the banjo tracks way down. If you look at it this way, a bird is just a new arrangement of an old tune. The dinosaur melody and the old genetic information are in there, but the sound is more contemporary. Of course, the better tune right here might be the old standard, the chicken song, "C-H-I-C-K-E-N, That Is the Way to Spell Chicken."

I have a chicken skeleton on my desk at the Museum of the Rockies. I have for decades kept a chicken skeleton at hand wherever I have worked, because it looks like a dinosaur, and I like being around dinosaur skeletons. Sometimes I look at it and turn it this way and that and think, If I could just grow these bones a little different, tilt this one way, that another, I'd have a dinosaur skeleton. Over the past few years I have been looking at that chicken skeleton more often and more intensely. As I've looked at the bones I have started thinking less about the bones and more about the underlying molecular processes that caused the bones to grow. And the more I've learned about evo-devo and looked at that chicken skeleton, the more reasonable an idea it has seemed. That skeleton started out as an embryo, a single cell, dividing and growing, the cells differentiating into different types. Chemical signals directed by DNA turned it away from the path of growth that would lead to a nonavian dinosaur, but it seemed highly likely that all the raw material, all the genetic information needed to grow a dinosaur, was in that embryo. How much, I wondered, would it take to redirect its growth so that it ended up looking like a dinosaur?

Experimentalists have already caused a chicken to grow teeth. Other researchers have chemically nudged chicken embryos to develop the different sorts of beaks that the famous Darwin's finches display.

Once I got the idea in my head that it could be done, I started talking to researchers who were truly grounded and fluent in the language, ideas, and techniques of both paleontology and molecular biology, like Hans Larsson at McGill. He was already working on what he called experimental atavisms as a way of understanding evolution. That is, he wanted to prompt a living creature to develop an ancestral trait.

The short version of the story is that I recruited Hans to drive the chicken/dinosaur express, at least part of the way. Hans is not growing a dinosaur, not yet. And none of the embryos in his experiments will hatch. But his research and my waking dream of having a chicken-sized dinosaur with teeth, a tail, and forelimbs instead of wings fit well together. So I supported research on getting that chicken embryo to express its inner tail. He has already discovered aspects of tail growth that tie this process to the very basic and early directions for the growth of any vertebrate, including humans. So the work may have unexpected value for some of the most common and devastating birth defects, those affecting the early growth of the spinal cord.

Researchers have shown how beaks can be modified by a change in one gene. If successful, Hans will show how the dinosaur's tail turned into the chicken's pygostyle and how to grow a chicken with a tail instead of a pygostyle. He will also have laid the groundwork for future research and, perhaps, for finally hatching that living example of how evolution works.

The scientific rewards of the process of learning how to rewind evolution would be enormous. It would be a remarkable demonstration of a direct link between molecular changes in the developmental process and large evolutionary changes in the shapes of animal bodies. We know quite a lot about the evolution of different forms from the fossil record. And we also know about specific changes in DNA that cause identifiable changes in body shape in the laboratory, among fruit flies in particular. But we are just now beginning to link molecular changes to large changes in the history of life, like the loss of a tail.

A nonavian dinosaur has not yet hatched from a chicken egg. This book is about how that became a goal I want to pursue, and how that pursuit is continuing, about how I and other scientists have tried in every way to travel into the past and bring it to life, how we began with pick and shovel, moved to CT scans and mass spectrometers, and have now arrived at the embryology lab.

The story starts with old-fashioned fossil hunting, with the hunt, the discovery, and the digging, always digging. There are two excavations here, both of the same *Tyrannosaurus* skeleton, a treasure of a find nicknamed B. rex, taken from the badlands of eastern Montana. The story continues with a second, microscopic excavation of the fossil bone itself by Mary Schweitzer, and the discovery of what seem to be fossilized remnants of sixty-eight-million-year-old blood vessels, still flexible; what may be the fossil remains of red blood cells and bone cells; and protein molecules, or parts of them, with unchanged chemical structure.

There comes a point, however, in studying dinosaurs and in this story, when fossil bones can yield no more. The search for more information about how dinosaurs evolved has to shift to the genes of living creatures. That is where the story takes a sharp turn, like the one the chicken's growing tail makes as it stops growing and turns into something else. And there it chronicles how and why we know that the history of evolution is written in the genes of modern dinosaurs, the birds, and how and why we can take a chicken egg that might have become part of an omelette or an Egg McMuffin, and convince it to turn into the kind of dinosaur we all recognize.

When we succeed, and I have no doubt that we will, and sooner rather than later, it will be another step in a long chain of attempts to re-create the past. The first public exhibition of dinosaurs was in England in 1854, at the Crystal Palace in Sydenham, five years before Darwin published his *Origin of Species*. It was groundbreaking, although the stances of the animals were wrong, putting dinosaurs that stood on their hind legs on all fours instead. Over time, our reconstructions of dinosaurs have become more sophisticated and more correct. We have changed the stances of dinosaurs in museums, learned that many of them were warm-blooded, that dinosaurs, not birds, were the first to have feathers, that some dinosaurs lived in colonies and cared for their young in nests, that many of them were much smarter and more agile than we'd ever imagined.

We have made accurate robots and 3-D reconstructions of the internal cavities of skulls. We have theorized about the colors and sounds and behavior of dinosaurs. And now, we can try to make a living dinosaur. This is a project that will outrage some people as a sacrilegious attempt to interfere with life, and be scoffed at by others as impossible, and by others as more showmanship than science. I don't have answers to these challenges, really, because the answers are not mine to provide. I have my ideas, my concerns, my own questions about the value and difficulties of such a project, but the story I have to tell is, like science itself, more about questions than answers, and the book is not a recipe or lecture. When we get to the point of hatching a dinosaur, it will be a decision that involves society as a whole, not just a few scientists in a laboratory. Most of all, this book is an invitation to an adventure. I can say how it begins, but all of us will have a say in how it ends.

1

HELL CREEK

TIME, SPACE, AND DIGGING TO THE PAST

"Scientific people," proceeded the Time Traveler, after the pause required for the proper assimilation of this, "know very well that Time is only a kind of Space."

—H. G. Wells, *The Time Machine*

To get to Hell Creek, you drive east from Bozeman. And back in time.

Bozeman is firmly located in the present. Twenty-five years ago it was a sleepy college town, where you could stop into a video poker bar and play pool while the Montana State University rodeo team celebrated at a nearby table. The town had coffee shops and college burger joints and creeping vegetarianism, but styles still lagged behind the coasts.

Things have changed. In 1980 Bozeman had a little more than twenty thousand people. Now it has about thirty-eight thousand. The nearby Gallatin, Madison, and Yellowstone Rivers have drawn fly fishing tourists, the Paradise Valley has attracted Hollywood royalty, and the foothills of the Rockies have become dotted with vacation homes of Angelenos. Early settlers in this migration, like Peter Fonda and Ted Turner, seem almost like natives now. High fashion may belong to the East and West Coasts. But as for outdoor chic—what's cool in mountain bikes, running shoes, and river sandals—Bozeman is at the forefront. It is not a city where you end up by chance. It is a destination.

You can't blame the new Montanans for coming here. They are drawn by the rivers and the mountains, the big sky east of the Rockies, and the rich fir forests west of the Rockies. And the fishing.

From a paleontologist's point of view the resulting change is something like deposition. Geological environments are of two sorts, depositional and erosional. At the extremes, a river delta deposits a lot of silt and builds up the land, while a mountain slope is constantly eroded by wind and rain. All of history is depositional, I suppose, in the sense that events accumulate, languages and cultures change, adding layer upon layer of human experience. In Bozeman the rate of cultural deposition is high, and increasing.

Espresso is now the drink of contemporary Montana. I'll wager that it's easier to get a triple skim latte around Bozeman than it is on the Upper West Side of Manhattan, at least if you don't want to get out of your vehicle. We have drive-up espresso stands, where the pickups and the Priuses idle in line together. Latte is the cowboy coffee of the twenty-first century.

Bozeman is just east of the Rockies, in the foothills of the mountains. To the south lie the deep, fir-banked canyons of the Gallatin River and the bison and elk of Yellowstone National Park. To the west is farmland cut by the Madison River. In July the river is filled with small drift boats, skippered by fishing guides, often college-educated northeasterners who practice the religion and business of trout in Montana in the summer and Argentina in the winter.

To the northwest lies the Blackfeet Reservation and, beyond that, Glacier National Park with its Going-to-the-Sun Road that takes you over the Continental Divide. To the northeast lies the high line, the towns that dot the rail line just beneath the Canadian border, towns like Cut Bank, and Shelby, where I grew up.

All of these routes have their attractions, but for time travel I recommend driving due east. Take Route 90, a modern four-lane highway that parallels the Yellowstone River, which itself was a highway of sorts in earlier times. The twenty-first century keeps you in its grip for a good 150 miles until you leave the highway at Billings, the state's largest city, with a population of close to one hundred thousand. Take Route 87 north, a flat, two-lane road through farm- and rangeland that leads to a crossroads marked by an enormous truck stop on the southeast corner. Look east down Route 200 to the heat-rippled horizon. That's where the past lies.

The next town is Winnett. It is small, population less than two hundred, hidden from the road, and deep enough into the all-but-empty western end of the Great Plains that there are no familiar fast food chains. There are, however, satellite dishes everywhere. And on the way into town a few years ago you could see small roadside signs made as part of a high-school antidrug campaign—unhappy reminders of modern life on the wide-open, economically depressed High Plains. "Violets are Blue, Roses are Red, If You Try Meth, You'll Be Dead."

The gap between Bozeman's prosperity and the economic situation of eastern Montana is significant. The state as a whole had more than 14 percent of the population under the poverty line in the 2000 census, which partly indicates the dire situation of the several large Indian reservations in the state. In Gallatin County, which includes Bozeman, the rate was 12.8 percent. In Garfield County, which does not include any reservation, it was 21.5 percent.

From Winnett east the road runs through open country for the next seventy-five miles. There are no towns. Other cars and trucks are few. The power lines are not always visible. The road runs straight to the horizon. Even traveling at eighty miles an hour, you can feel the distance and the emptiness. In a car with a tankful of gas it is exhilarating. On foot it can be oppressive. If the space seems endless now, what did it seem two hundred or one hundred years ago?

The road cuts through the Missouri Breaks, the same eroding badlands that Lewis and Clark passed through, that Sitting Bull knew, that were here when humans came across to North America from Asia and began to hunt mastodons. Well, almost the same. The short-grass prairie has changed. Sagebrush, a staple of old Western movies, is everywhere, the result of overgrazing by cattle and sheep as far back as the early 1900s. When the buffalo lived on the plains in the millions, the land was still dry and sparsely vegetated. But the short grass dominated. The land was always harsh and never drew agricultural settlements of Indians. It did support abundant wildlife, although the buffalo were not here in their endless herds when the first humans hunted here. One school of thought is that the buffalo and the Great Plains evolved as the first Americans burned the prairie to keep it open.

The Musselshell River marks your passage into Garfield County, and you begin to see exposed rocks, many around sixty-five million years old. You now occupy several slices of time. You are in the present, driving back to what feels like the early twentieth century, or the eighteenth or seventeenth. But all around you the torn earth reveals the deep past, open for investigation.

THE HELL CREEK FORMATION

The rocks of what geologists call the Hell Creek Formation are what you see from the road. The formation was named in 1907 by Barnum Brown, a paleontologist from the American Museum of Natural History, based on rock layers near Hell Creek. It is a mixture of sandstone and siltstone that is exposed in Montana, Wyoming, and North and South Dakota. The rock beds that make up the formation vary in thickness, or depth, from 170 meters in Garfield County, where Barnum Brown first identified it, to 40 meters in McCone County, Montana. The formation preserves somewhere between 1.3 and 2.5 million years, depending on different interpretations.

These are not just any million or so years. These rock beds are made of sediment deposited at the end of the age of dinosaurs. The Hell Creek Formation is probably the best record anywhere on earth of terrestrial life at the end of the Cretaceous era (210 million to 65 million years ago), just before and right at the mass extinction that wiped out about 35 percent of all species, including the dinosaurs.

The end of the Cretaceous era is marked in much of Garfield County at the top of the Hell Creek Formation by a dark line in paler rock. The line is coal, deposited just after the mass extinction, and marks the beginning, in this location, of the Fort Union Formation. This, at least, is a rule of thumb. Elsewhere in the world chemical markers give a very specific and clear way to mark the end of the Cretaceous. In eastern Montana, however, these chemical markers are not always present, so the process of refining the date at which the Cretaceous ended and the mass extinction occurred continues.

Nonetheless, the dark line, called the Z-coal by some researchers, is a stark reminder of the vertical progress of deep time. The abstract numbers in the millions may be hard to comprehend, but the rock, in varying color, marking the agonizingly slow accumulation of silt from overflowing streams, sand from beaches, or rotting plants transformed to coal, is there to see and touch.

The formation is rich with evidence of ancient life. Here, in this rock, there are dinosaurs, like the fossils of *Tyrannosaurus* that Barnum Brown found in 1902, although he didn't know they were *T. rex*. Actually, *T. rex* had not yet been named. There are huge bone beds of *Edmontosaurus,* a duck-billed dinosaur, found at the boundary of North and South Dakota, apparently all killed in a storm surge from the shallow inland sea that covered the center of the continent.

There are numerous other dinosaurs from the end of the Cretaceous, *Triceratops* and the boneheaded pachycephalosaurs. And then, there are none. Ever again, anywhere in the world. Deposition of sediment that turned to rock continued in this part of North America. And there are plenty of fossils in rock beds from later dates that document other passages in the history of life, like the rise of the mammals. There are bird fossils. But the nonavian dinosaurs are gone. Forever. Locked in deep time, while the planet piles up rock, erodes mountains, and moves continents.

The extraordinary fact about Garfield County is that it presents so clearly one particular chapter in the history of the planet. The sixty-five-million-year-old rocks of the Hell Creek Formation are near or right on the surface, ready for further excavation. This formation has yielded railroad cars full of fossils, including the *Tyrannosaurus* skeletons in the American Museum of Natural History and the Smithsonian Institution, and, of course, the fossil skeleton of *T. rex* that sets this story in motion, a find that we nicknamed B. rex, *B* for Bob Harmon, who discovered it.

Here in Garfield County is the section of the Hell Creek Formation that is best known and most researched. This is a part of Montana that has a bigger sky than the rest of it, if that's possible. It has only a thin layer of cultural deposition by modern

Americans, or indeed by any other human beings who have passed through in the past thirteen thousand years or so.

Garfield County covers more than three million acres, almost five thousand square miles. It is roughly the size of Connecticut with a population of about twelve hundred, although it would not be easy to find them all. At the population density of Garfield County, four square miles per person, the island of Manhattan would be home to five people.

Jordan, the only town in the county, population about 350 and dropping, is where you end up after the seventy-five-mile drive from Winnett. Jordan is known among fossil hunters and fishermen as the last stop before they continue on their quests, often by heading down the same dirt road toward Hell Creek State Park. The fishermen are after the walleye in Fort Peck Lake. The land surrounding the lake is part of the Hell Creek Formation and yields fossils of mammals, reptiles, shellfish, and plants as well as dinosaurs.

Visitors to Jordan tend to remember the Hell Creek Bar. It has ice cold beer, chicken in a basket, and a long wooden bar at which ranchers, fossil hunters, fishermen, and, on occasion, reporters mingle. Jordan gets more press than you might imagine for a town its size. Part of it is for the dinosaur fossils. It has been a town of note to geologists, paleontologists, and their audience since the days of Barnum Brown at least.

But it has some more recent claims to fame, not all savory.

In 1996 Jordan was the site of a small rebellion against the United States on the part of a group called the Freemen, a white supremacist militia that specialized in bank fraud. The group were followers of Christian Identity, a creed that holds that white people were descended from Adam and Eve and Jews and people of color from Satan and Eve. They did not recognize the United States, and set up their own banking system, which consisted largely of a computerized forging operation that brought in a total of nearly $2 million, according to the government.

The Freemen, led by LeRoy Schweitzer, conducted courses in forging financial documents, filed so-called "liens" against government officials, and wanted to return the government of the country to white males and fight the international banking conspiracy, led, of course, by Jews.

By the summer of 1996 they had alienated all their neighbors in Garfield County, near Jordan, where they had set up a compound at the 960-acre Clark ranch. The FBI arrived in Jordan after arresting several of the leaders who were not at the ranch. The siege of an extremist camp might have turned ugly, but the death of seventy people at Waco, Texas, three years earlier in a raid on the Branch Davidian complex there was still fresh enough that what followed was a long, nonviolent siege—eighty-one days.

The siege stayed in the news intermittently, although there was only one death. An FBI agent died in an automobile accident when he ran off one of the unpaved roads in the area. The siege ended without any dramatic confrontations. Eventually the Freemen all left the ranch and many were arrested. Some disrupted their trials by refusing to recognize the court's authority. The Freemen faced a variety of charges, including bank fraud, mail fraud, and armed robbery. Eight of the men received twelve-to-eighteen-year sentences. Others received lesser sentences and LeRoy Schweitzer, the leader, was sentenced to twenty-two and a half years and is still in prison.

The Freemen standoff caused painful divides in families, between brothers and sisters, and parents and children. Many of the people who lived in and around Jordan thought of the Freemen as a cult. This was before Montana State set up a dig at Hell Creek, but our research staff was touched by the events, as were many people in Montana, who saw friends, neighbors, and relatives somehow drawn into the Freemen. Mary Schweitzer, who, as I mentioned in the introduction, led the research on the fossil bone tissue of B. rex, was connected by marriage to LeRoy Schweitzer, the leader of the group. He is the brother of her ex-husband.

That was a difficult summer for Montana in other ways. Ted Kaczynski was found living in a cabin in Lincoln, on the western side of the Rockies. Kaczynski, the notorious Unabomber, was bigger news than the Freemen. He had sent bombs through the mail to people he thought were responsible for the ruin of modern society by technology. Over the course of about two decades he killed three people and wounded twenty-two. He was eventually identified by his own brother and he turned out to be an academic, with a Ph.D. in mathematics, who had gradually detached himself from society and embarked on a violent crusade.

It's an old rhetorical flourish to tie politics to the land, and often false. The Freemen had no support from the population around Jordan, partly because they didn't do any honest work. But it is true that eastern Montana is extreme even in a state given to extremes, in landscapes, weather, and history. And as life gets easier in other parts of the country, it just seems to get harder to make a go of it in Garfield County. The land can be harsh to the point of desolation. In the summer, dry mudstone flats bake in 120-degree heat and drinking a gallon or two of water a day to keep hydrated becomes a matter of survival. In the winter the wind rages at 40 below zero. The end of nature may have arrived in principle, but in a place where cell phone signals often disappear, the ancient hazards have not lost their power.

The Missouri River is the county's northern border, in the form of Fort Peck Lake, 134 miles long, rich with fish and often shrunken by drought. The lake is a product of the Depression-era Fort Peck Dam, built from 1933 to 1937 to provide power and jobs. Ten thousand people worked on the dam, just over the line in McCone County. Since the dam was finished, nothing else has brought people to this part of Montana in those numbers.

The geological past seems to dominate the human story here the way the weather can overwhelm philosophical musings about the planet. The earliest ranches here are only a few generations old. The hold of people on the badlands feels tenuous. The fossil hunters for the great New York and Washington museums, who arrived at about the same time as settlers, struck it rich, so to speak, filling the halls of natural history museums with their discoveries. The ranchers have hit no jackpots.

But from the Indians to the Freemen this is a thin history. This is not a depositional environment, for human culture or rock. The badlands erode and the people leave. The past remains. About a million acres of Garfield County are occupied by the Charles M. Russell National Wildlife Refuge. The rest is a mix of public land overseen by the federal Bureau of Land Management and private ranches. All of it is badlands, crisscrossed with a grid of gravel roads that organize what would otherwise feel truly desolate. The landscape can be beautiful when deep shadow and blinding light fragment the geometry of the gullies and bluffs. It is raw and unpolished, by people or nature.

Leaving Jordan for fossil hunting, you take gravel, or plain dirt roads. You establish a camp, with all the amenities of the modern age that you can muster. Electronics are easier than plumbing. You can set up a satellite dish for broadband computer access. You can even webcast from camp. But outhouses are the rule. When you leave camp for the day to prospect, or dig, you leave the outhouses behind.

Once you are in the field and you are excavating the thighbone of a tyrannosaur, you have gone millions of years back in time. You stand, with parched throat and sweat-soaked shirt, without shade or water, in the present. But you are really back in the Cretaceous, as the fossils and geology show, in a marshy river delta on the shores of a shallow sea that bisected the North American continent from the Arctic to the Gulf of Mexico.

And you can see all this in the rocks. This readability of the earth's past is a wonder many of us take for granted, speaking of the Cretaceous or the Jurassic, offhandedly describing the animals that lived then, the environment, the weather, temperature, and the arrangement of the continents. But our ability to look back at the past, to re-create it with some confidence, at least in broad strokes, is a wonder that far surpasses the tales told by any religion.

We are now multiplying this wonder, adding the tales told by molecular fossils, and the history of life's evolution written in the DNA of living animals. What we find gives us new understanding of the past, and new ways to try to reconstruct it. And some of the most astonishing finds, as has been true for at least a century, came from the Hell Creek Formation in Garfield County.

THE DINOSAURS

To understand the place of the Hell Creek Formation in the history of life on earth it is necessary to step back a bit, perhaps not to the origin of the planet four and a half billion years ago, but at least to the beginning of the dinosaurs' reign. By the time of the nonavian dinosaurs' extinction, this extraordinary group of animals had already had quite a successful run.

All of life's history is of a piece, so it is awkward to step in at any given point. At the time of the origin of dinosaurs 225 million years ago, during the Triassic period, the big stories in the evolution of life were finished. The first hints of life appeared about 3.8 billion years before the present, and the first, most primitive soft-bodied animals not until about 650 million years ago.

This was just before an enormous blooming of animal forms in the Cambrian period that filled the seas with wriggling, swimming, voracious life. Vertebrates evolved. And fish. About 360 million years ago animals and insects began to colonize the land. That was when the first tetrapod appeared. Tetrapods, four-limbed vertebrates, are, or should be, near and dear to us, since we ourselves are tetrapods. We are one variation on the tetrapod theme that has been sung by natural selection for several hundred million years. We may not be able to claim the antiquity of jellyfish, the variety of insects, the biomass of bacteria. But still, we have reason to take some satisfaction in how successful this basic body plan has been. Reptiles are one highly successful variation, and among the reptiles the much-beloved dinosaurs are unarguably memorable.

We cannot point to a particular fossil and say, here, this is the ancestor of the dinosaurs. In fact, evolutionary biologists have changed how they group animals and categorize descent from ancestral forms, since the familiar and easily understood tree-of-life diagram was developed to grace textbooks and magazine articles. Biologists no longer look for "the" ancestor. Instead they concentrate on shared characteristics that define one group and the new characteristics that appear in the course of evolution to define a new group.

These new traits are called derived characteristics. And the groups are called clades. A diagram of the course of evolution using clades is a cladogram and it is somewhat similar to the old tree. Single-celled organisms are at the start of things, fish appear before amphibians, and both before reptiles. Humans, of course, branch out very late from our mammalian and primate origins.

But cladograms don't pin down ancestry to one species or genus. A cladogram shows, for example, that from the vertebrates

Jack Horner looks over eroded badlands near Jordan, Montana. The boundary that marks the end of the nonavian dinosaurs and the top of the Hell Creek Formation is a dark line about two-thirds of the way up the hill, just left of center.

new clades have evolved that have all the shared characteristics of vertebrates plus some new, derived characteristics. For instance, all vertebrates have a backbone. A clade that evolved from the vertebrates, like the mammals, shares the backbone and vertebrate body plan, with eyes and mouth at the front and a digestive system that goes from front to back, among other details.

But the mammals have derived characteristics like mammary glands and fur that we don't share with other vertebrates. By looking at more detailed skeletal characters, we can see that the ancestors of the mammals branched off from the reptiles before the evolution of the dinosaurs. In fact, the mammals and the dinosaurs appeared around the same time.

Scientists agree that all the dinosaurs come from one ancestral source. The dinosaurs themselves are in two groups, the Ornithischia and Saurischia, based on the structure of their hips, but these groups have a common dinosaur ancestor. We can't say what that specific ancestor is, or what creature immediately preceded the dinosaurs, but we do think it was something like *Lagosuchus,* a reptile that was less than a meter long and walked on its two hind legs. The first dinosaurs were also bipedal, and the four-legged stance of familiar animals like *Triceratops* and *Brontosaurus (Apatosaurus)* evolved later.

A mass extinction killed the dinosaurs and it may have been a mass extinction that gave them their start. There were two extinctions in the Triassic, one around 245 million years ago, perhaps caused by an asteroid hitting the earth, and one 205 million years ago. After the first, many of the ancestors of mammals and dinosaurs disappeared, leaving some good opportunities. The dinosaurs took over. The mammals stayed in the background for 145 million years until the next mass extinction did away with the dinosaurs and once again offered abundant unfilled ecological niches.

The time of the dinosaurs' origin was a good one for land animals in the sense that all the modern-day continents were united into one landmass called Pangaea. These continents had previously shifted and drifted apart in various combinations, and once before had united in a supercontinent called Gondwanaland. That broke up, and so, eventually, did Pangaea. The continents continued to drift as the dinosaurs evolved.

Dinosaur species appeared and disappeared over the next 140 million years, taking on the numerous forms that fossils have preserved, the gigantic sauropods, like *Apatosaurus,* carnivores like *Allosaurus* and *T. rex,* the plant eating duck-billed dinosaurs, small colonial nesters such as *Protoceratops,* agile small hunters like *Deinonychus* and *Velociraptor.* Along the way, birds emerged.

By the time of the latest Cretaceous, the period preserved in the Hell Creek rocks, Antarctica, Australia, and South America had separated from the unified landmass and were not connected to other continents. But Africa was still attached to Europe and northern land bridges connected North America, Europe, and Asia.

The sauropods were gone, and among the dominant land animals was *Tyrannosaurus rex.* Duck-billed dinosaurs abounded, as did *Triceratops* and other dinosaurs like the *Pachycephalosaurus.* The late Cretaceous was a time of mountain building in North America. The rising mountains that we now call the Rockies were being eroded as fast as they grew and were drained by rivers and streams that dumped sediment on the plains during floods, burying and preserving dinosaur bones.

The rivers and streams in what is now Montana flowed into the inland sea, which expanded and contracted over millions of years. By the time of the latest Cretaceous, when the Hell Creek sediments were deposited, the sea had retreated and the section of the formation in Garfield County, from which paleontologists have drawn so many fossils, was a river delta, with winding channels, and both land and water habitats, near the coast of the inland sea.

The vegetation was thick and varied, including ferns, conifers, and flowering trees. These are known both from fossils and the microscopic analysis of pollen grains in the rock. Herds of dinosaurs fed on the lush plant life and were preyed on by packs of smaller, hunting dinosaurs like *Troodon*. Lizardlike predatory mosasaurs and long-necked plesiosaurs swam in the inland sea. Mollusks were present in the seas, as well as ponds and rivers, which played host to fish, amphibians, crocodilians, and turtles; all of these survived and continued to thrive when the dinosaurs disappeared. There were land, shore, and diving birds and numerous mammals, primitive relatives of today's egg-laying platypus and echidna, marsupials, and placental mammals. Some of these would have run up and down the trees like squirrels. Others would have lived on the ground, perhaps in burrows.

One day, while life hummed, chewed, killed, and died, as usual in this tropical environment, a meteor ten to fifteen kilometers in diameter—six to nine miles—entered the earth's atmosphere, headed for the sea near the Yucatán Peninsula at about seventy thousand miles per hour (thirty-two thousand meters per second). It was traveling at a thirty-degree angle when it struck. The meteorite vaporized, as did its target. One estimate is that twelve thousand cubic miles of debris were sent into the atmosphere. And the energy released has been estimated at one hundred million megatons. The blast at Hiroshima is estimated at fifteen kilotons. In other words, the explosive energy released when the meteorite hit the earth was the equivalent of 6.6 billion atomic bombs like the one dropped on Hiroshima exploding simultaneously.

The result was a worldwide disaster, although it is hard to pin down the exact effects. The impact coincided with a mass extinction that wiped out 35 percent of the species on earth, including all of the nonavian dinosaurs. For a brief time the foundation of the world ocean ecology, the community of microorganisms that harvest the energy in sunlight, was devastated.

The evidence for this event was first published in 1980. Walter Alvarez, a geologist, had been in Italy studying the rate of accumulation of cosmic dust in geological strata as a way of dating them independent of fossils or other methods, when he found that right at the K/T (Cretaceous-Tertiary) boundary, there was much more iridium than in any of the other strata. He was near the town of Gubbio, and in the rocks he was studying, the K/T boundary was marked by a layer of clay. Below it were fossils of microorganisms of the Cretaceous. And above it were fossils of different microorganisms, from the Tertiary.

This boundary is visible in other formations around the world and was long known to mark a great extinction of the dinosaurs and many other forms of life. There are at least two other mass extinctions, one at the end of the Triassic, about 205 million years ago, and another one, the most significant known so far, at the end of the Permian, 250 million years ago. Ninety-five percent of all species in existence then died out.

Only in recent years have paleontologists and evolutionary biologists come to recognize the importance of mass extinctions in evolution. These events brought chaos and destruction to the planet, and opportunity. New forms of life evolved rapidly to occupy niches in the environment left open by the disaster.

But explanations for these extinctions have been hard to come by. There has been no end of argument about the extinction of the dinosaurs. So the discovery of this highly unusual concentration of iridium exactly at the time of a mass extinction was intriguing. One of the events that would produce such a spike would be the impact of an asteroid hitting the earth. Iridium is common in asteroids but not in the earth's crust and an asteroid of sufficient size—about ten kilometers in diameter, Alvarez estimated—would do the trick. He worked with his father, Luis Alvarez, a Nobel Prize–winning physicist, and two geochemists, Helen Michel and Frank Asaro, all at the University of California at Berkeley. Michel and Asaro had found iridium spikes at two other sites marking the end of the Cretaceous—in Denmark and New Zealand.

A great scientific debate continued as many more iridium spikes in a clay layer of the same age were found, including in some parts of the Hell Creek Formation. But there were, and are, many puzzles about why some animals went extinct and others did not. And a question remained, for a time, about where the evidence of such a collision was. This was the kind of impact that would leave a mark.

It wasn't until 1990 that seafloor cores drilled in the Gulf of Mexico showed quartz that had been transformed by an impact of the sort an asteroid would cause. And it lay underwater in an area near the town of Chicxulub that had been spotted a decade earlier as a potential impact crater. That first claim did not attract scientific attention, but the "shocked quartz" did. Other evidence accumulated—glass deposits in Haiti and sand in Montana, blown from the crater.

What exactly the asteroid did to the global environment is not known, and explanations still abound for the mechanisms of extinction, but what is clear is that there was a massive and immediate global extinction after an asteroid impact of literally unimaginable proportions. And the best place to see the fossil killing field, or iridium layer, and the terrestrial life before and after it is the Hell Creek Formation.

A year and a half ago astronomers identified what may have been the source of the Chicxulub meteorite, a collision in the asteroid belt that occurred 160 million years ago in a group of asteroids known as the Baptistina family. As was pointed out several times in the news of the discovery, this would mean that the Chicxulub asteroid was set on its course about 100 million years before it hit, in the middle of the Jurassic age.

Any modern human with even a hint of pessimism about the future of the human race has to have some sympathy for the creatures of the Hell Creek ecosystem, completely unaware that many of them were about to disappear forever. Of course, some of them were about to get their big chance— mammals, for instance. They were small, perhaps able to survive on a variety of foods, including leftovers from the extinctions, and not dependent only on photosynthetic plants, which seem to have had a hard time.

Evolution driven by catastrophe is not what Darwin had envisioned. He, and many of those who came after him, saw natural selection acting gradually, preserving, or selecting for, the traits of animals that left more offspring. That would still happen in a catastrophe, of course, but it would not be a honing of traits best suited to particular niches. For a time chaos would serve those organisms that could thrive in a wide range of environmental conditions, and at first there would be little selection of the sort Darwin imagined, because as the damaged planet recovered, there would be plenty of room for the fit and unfit to prosper, as long as they weren't too ecologically picky.

Think of the Permian extinction, for instance. Before the extinction, in which 95 percent of species were wiped out, many changes in behavior or form prompted by mutations in genes would have been lost, because organisms had particular niches that they had evolved to exploit and too much deviation would probably diminish their fitness. After the extinction, however, refinements in exploiting one kind of environment might be nothing compared to a fast reproductive rate and an ability to eat anything and everything. If every niche were opened by the extinctions, the world would be welcoming to all sorts of mutations.

Nature had suddenly become a kind of Wild West, far less picky about whom it welcomed. It was a new frontier of sorts, something like western North America when it was being taken over by European-Americans. In the West manners were much more varied than in the East or in England. Behavior that would not be tolerated in a stratified society in which all the niches were filled was tolerated, or accepted, in a land that offered all sorts of opportunities, once the original inhabitants were gotten rid of. In the American West social mobility was great. After a mass extinction evolutionary mobility was greatly enhanced.

About 35 percent of the existing species were wiped out in the K/T extinction, a mere interruption compared to the catastrophe at the end of the Permian. The consensus in science seems to be that the asteroid impact was the primary cause of the extinction. And the meteor crash is so astonishing in its destructive power that it tends to obscure the time before and after it landed off the Yucatán.

But the period before the crash is fascinating. If we go back ten million years before the meteor hit, dinosaur diversity was at a peak. This is a time when the Judith River Formation in Montana was laid down. And this formation is rich in the numbers of different species. But then, when we turn to the Hell Creek Formation, ten million years later, we find many fossils, but far fewer different species. And the more we study the fossils, the fewer species we find.

Recently, some of the species of the Hell Creek Formation have gone extinct. In this case, the cause is paleontological. As we understand more about the growth of dinosaurs, we find that some specimens that we thought were different species are just different ages of the same species. We used to think that *Dracorex, Pachycephalosaurus,* and *Stygimoloch* were three different species of dome-headed dinosaurs that were found in the Hell Creek Formation. Now we have found that Dracorex and Stygimoloch are juvenile stages of *Pachycephalosaurus.*

Many other species actually disappeared, rather than being reidentified. In fact, the biggest drop in diversity in the 140-million-year history of the dinosaurs occurred in the 10 million years before that meteor crashed. Something took out a lot of dinosaur species before the meteor finished the job. That is what interests me more than the mass extinction of the meteor crash, perhaps because the mechanism is so unknown and hard to understand. A meteor crash is, at heart, simple. Not that it isn't a challenge to figure out exactly what kind of havoc the meteor wreaked, but it's pretty clear that it caused a world of trouble. We don't know anything about why dinosaur species disappeared at a rapid rate in 10 million years before the extinction.

The unanswered questions may only increase the sense of awe that comes from standing in the Hell Creek Formation and seeing the coal that marks the end of the Cretaceous. All around you are elements of a fossil snapshot of the world just before a catastrophic event. Prospecting for fossils doesn't just produce the discovery of new species. Each fossil is a pixel in the increasingly detailed image we have of the moment before the extinction. That is the time frame you occupy in Garfield County: the moment before the end.

Or you can occupy the present, at least the present of Garfield County, which is a bit of a dislocation from, say, the present of Berkeley, or New York, or Washington. Fortunately, the ageless feeling of the rangeland in Garfield, the sparse human population, the quiet at night and the wide open sky provide a cushion against the gap in time, so that it does not feel so odd to be straddling the eons. But one thing that the rocks of Garfield County do not offer is any sense of what happened between then and now. Life did not cease on this patch of the planet's surface with the extinction. Sixty-five percent of species survived, some prospered, and many new ones emerged. But no record of these events was preserved here. The deposits continue for a few more million years after the K/T boundary. Some of that time is preserved in deposits in and around Jordan that look much like the Hell Creek deposits except that they are tan rather than gray, there is more coal, and it seems to have been swampier than before the extinction.

The Paleocene lasted for another ten million years, and some of that can be seen in Garfield County badlands. But not much later. Life continued in the area, but we have no sedimentary rock from that time in that location. The planet's surface is a patchwork of different time exposures, like a canvas painted over many times, with different works showing through at different spots where the paint is thicker or more has been scraped off by curious art historians.

We have to turn to other fossil records, of which there are many, for an idea of how this part of the planet made it from then to now. If only for the sake of context, it is worth stopping to fill in a few of the blanks.

SINCE THE DINOSAURS

What has happened to North America since that time? The answer is: almost everything. The continent, and the world, went through geological and climactic upheavals. Mammals began to radiate into forms that seem outlandish today. In their range of shape and behavior they challenge the dinosaurs, although the dinosaurs get most of the press, perhaps because there are so many mammals around now, such as humans.

If one is tempted to think of the mammals as a poor sequel to the dinosaurs, it's worth remembering that they lived through the entire age of the dinosaurs as well. What happened after the extinction of the nonavian dinosaurs was simply that they became the dominant land animals, as they are today. The impact of the meteorite was a crisis, but a manageable one for life on the planet. As for the rock we all live on, the Rockies continued to thrust upward, other mountain ranges of the West emerged, seas disappeared and reappeared in the center of North America.

In what is now the High Plains of eastern Montana, the conditions were junglelike as mammals began to radiate. The emergence of these creatures and their radiation into so many forms is as vivid an indication of how evolution proceeds as was the era of the dinosaurs, or the evolution of birds, which coincided with the mammalian explosion. Every shape and size emerged among mammals, in configurations that we are familiar with—extremely large herbivores, like the uintatheres; predators with flesh-cutting teeth; small mouse- and shrew-like animals; the swift (deer- and antelope-like creatures); and the slow (great sloths).

All of these widely divergent creatures were variations on the standard mammalian model. All were furry tetrapods with five-digit hands and feet, hearts, lungs, and brains. Those brains were protected by the ancient skull structure that had been around before anyone had thought of fur. Changes in size are easy to understand occurring quickly. A different regulation of growth, a few genes turned on and off at different times, or producing more or less of the regulatory proteins, and the tiny mammal would roar, or bellow. Shape inevitably changed with size. Teeth changed with diet. Depending on food availability, stomachs and metabolisms changed. But over the course of sixty-five million years the fundamental mammal has stayed visible.

It is too long a time to track every change that occurred in the continental interior, or worldwide, where one branch of mammals, the primates, were developing bigger and bigger brains and new behaviors. Hominins first appeared about six million years ago, according to current thinking, shortly after the last common ancestor of humans and chimpanzees, probably six to eight million years ago. The succession of hominins that led to humans is long and not entirely clear. But we do know that not until about fifty thousand years ago did physically and behaviorally modern humans leave Africa. They quickly spread around the world.

When humans first came to the Americas is a matter of some dispute. The first undisputed evidence of their arrival puts the colonization of the open continent at a little over thirteen thousand years ago. These were the so-called Clovis people, who used distinctive stone points to hunt mammoth and other large animals. Stone points of this type were first found at Clovis, New Mexico. Humans may have arrived earlier, however, as some sites, like one in Monte Verde, Chile, have materials that date to more than fourteen thousand years ago and show evidence of settled rather than nomadic life. That would seem to suggest that these people came to the New World from Asia earlier than the Clovis hunters.

The first culture to be widely represented in North America, however, is that of the Clovis hunters, who quickly spread across the continent. They came at a time when the last glaciers were receding, and a passable overland route from Asia existed where the Bering Strait is now. If other humans had come to North and South America earlier, they did not leave a mark on the environment that we have found.

The Clovis people encountered animals that we have never seen, the last of the great mammals of the ice ages. Mammoths and mastodons were common. Huge short-faced bears, bigger than grizzlies, may have been present in smaller numbers. Long-horned bison were plentiful. There were at least some ground sloths surviving, as well as tapirs, a giant beaver, horses, and other animals.

All of these—the Ice Age megafauna—disappeared rapidly right at the time when the Clovis people hunted their way east and south. For some time these hunters have been viewed as the human equivalent of a Chicxulub meteor crash, a destructive wave of humanity sweeping east in a way that strangely prefigures the European sweep westward thousands of years later. A more recent view is that a change in climate played a role in the extinctions. Tim Flannery, in his excellent book *The Eternal Frontier*, still argues for the rapid extinction of large mammals by the Clovis people, who, he points out, left no art that has been found and seemed to live in more rudimentary shelters than did the earlier Asian cultures from which they came. Flannery suggests that they put all their effort into the beautiful and deadly Clovis points. He estimates that they accomplished the extinction of North America's ice age animals in three hundred years.

Whether or not he is right, it is clear that thirteen thousand years ago, the time when the Clovis people arrived, is the time when the animal ecology of North America changed. Brown bears, moose, and elk migrated from Asia. Gray wolves, a global Arctic species, replaced the larger dire wolf. The High Plains began to look something like they do today. Climate may have changed the vegetation, however, and the great herds of smaller, short-horned bison did not appear until the long-horned bison were extinct. The modern bison migrated from Alaska, and there is evidence that by twelve thousand years ago they were being driven to their deaths in herds.

The ancient American natives may have helped create the vast herds of bison that Europeans encountered, Flannery suggests, by their use of fire to create the kind of prairie that favored the growth of these animals in the millions.

The badlands of the Hell Creek Formation in Garfield County would have been the same, geologically, for the Clovis people as they are for us. They would have faced the same harsh environment we do when we hunt fossils. But the world around them would have been profoundly different. For ten thousand years the bison would have increased and the hunting cultures that depended on them would have visited the High Plains. One remarkable find near the town of Wilsall, about forty miles northeast of Bozeman, produced a spectacular accumulation of Clovis points from eleven thousand years ago that seemed to have been buried with a child of about eighteen months. The Anzick site, named after the owners of the land, also produced the bones of a second child six to eight years old. That child, however, died about two thousand years later, according to radiocarbon dating. Cultures changed during that time, but slowly compared to our own headlong rush. From New York, Los Angeles, or Bozeman for that matter, two thousand years of the Stone Age seem like a dreamlike stasis in which the land must have seemed eternal. From a camp in the badlands that same sense of being lost in time that moves at the pace of geology is almost reachable. But then the laptops come out at the end of the day, and with our dish antenna we connect to the world and to the impatience that speed brings. What would a Clovis hunter have made, not of our machines, but of our intolerance for delays of tens of seconds?

For thousands of years the people who hunted and quarried stone in Montana did so on foot, driving buffalo off cliffs to their deaths and setting up camps to butcher them where they fell. There are bluffs in Montana and other parts of the High Plains—buffalo jumps—with accumulations of buffalo bones at the base that indicate use of the same place to drive buffalo to their deaths over hundreds or thousands of years.

It was the arrival of Europeans, first the Spanish with the horse, and later the press of French and English westward in the United States and Canada, that changed everything. By 1492 the continent was populous and settled by many different Indian groups with distinctive languages and cultures. Many were agriculturalists, although not the Blackfeet or, later, the Sioux and Crow of Montana.

When the horse came, brought by the Spanish, life on the plains began to change, and the familiar culture of the Plains Indians, warriors on horseback in elaborate costume, began to emerge. Coronado visited the Wichita Indians in 1541. In the 1600s, fur trappers first introduced guns to eastern tribes. As white men pushed west, displaced and newly armed Indian tribes themselves pushed west. The historian Alvin Josephy writes that "by the 1740s, horses were possessed by almost every Plains tribe in both the eastern and western sections of the plains and as far north as Canada's Saskatchewan River Basin."

By the end of the eighteenth century, Josephy writes, just when Lewis and Clark were traveling through the plains, what we know from movies as the culture of the Plains was in full flower, with the war bonnets, lances, extraordinary horsemanship, and transportable tipi villages. These Indians were not farmers, they were hunters and fighters. The westernmost agricultural settlements were villages on the banks of the Missouri in the Dakotas.

Lewis and Clark entered what is now Montana in the spring of 1805. In the fall of 1804 the Lewis and Clark expedition stopped at a Mandan village in what is now North Dakota. The Mandans were agriculturalists and traders. In the spring of 1805 they left the Mandan villages. With Sacajawea, the wife of a French Canadian fur trapper, Toussaint Charbonneau, they journeyed up the Missouri in six canoes and two pirogues. Stephen E. Ambrose writes in *Undaunted Courage,* his book about the Lewis and Clark expedition, that eight days into the trip "the expedition passed the farthest point upstream on the Missouri known by Lewis to have been reached by white men." Lewis and Clark then passed above the badlands of present-day Garfield County with little incident. North of the river lay the territory of the Blackfeet, whom they did not see. Of the land to the south Lewis wrote on May seventeenth, just a few days before the party reached the Musselshell River, "the great number of large beds of streams perfectly dry which we daily pass indicate a country but badly watered, which I fear is the case with the country through which we have been passing for the last fifteen or twenty days."

In the same entry he notes that Clark was almost bitten by a rattlesnake, thus picking out one of the salient characteristics of the land. Two days later Clark noted in his journal one of many encounters with a grizzly bear—they killed it, as they usually did—and a violent confrontation with a less predictable adversary. "Capt. Lewis's dog was badly bitten by a wounded beaver and was near bleeding to death." This incident has been little commented on by historians, but it speaks to the determination of beavers.

The expedition did not linger in this area or the Missouri Breaks farther west. They continued on to the headwaters of the Missouri, across the Rocky Mountains and to the Pacific. On the way back in 1806, Clark and Lewis split up, and Clark and his party passed to the south of the Hell Creek area, traveling the Yellowstone River.

Lewis and Clark marked another major step in the advance of European Americans. Although their passage itself caused little damage, they were the tip of the spear, and behind them came the advance guard of white expansion—mountain men and traders. Jim Bridger, a mountain man celebrated for his wild and woolly ways, after whom the Bridger Wilderness in Montana is named, did not strike out into completely unknown territory in the pure entrepreneurial spirit that Americans prize so much. He first saw the land as a member of Lewis and Clark's government-sponsored expedition. During the ensuing decades, few people, Indian or white, would have spent much time in the badlands of eastern Montana, except to hunt buffalo. And buffalo hunting would not reach the level of wholesale slaughter until the coming of the railroads, after the Civil War.

The government took a hand in bringing the West under the heel of civilization during the Indian Wars. The most famous battle of this campaign against the aboriginal inhabitants of the West was probably the fight at the Little Big Horn on June 26, 1876, eight days before the centennial Independence Day celebration. As every schoolchild knows, or used to know, the battle was won by the Indians, who had no chance at all in the larger war.

The Little Big Horn is southwest of Garfield County. In a recent book, *Hell Creek, Montana: America's Key to the Prehistoric Past*, Lowell Dingus, a research associate at the American Museum of Natural History, recounts the pursuit of Sitting Bull and four hundred Indians after the battle by Colonel Nelson A. Miles. The colonel followed Sitting Bull, through the Hell Creek badlands in the fall of 1876, struggling with rough terrain and impending winter. In November temperatures were already dropping to twelve below zero.

Sitting Bull was apparently attracted to the Hell Creek area because the hunting was good. The buffalo were still there, feeding on the short grass of the western prairie. Neither Sitting Bull nor the buffalo fared well in the end. Sitting Bull eluded Miles and crossed over into Canada. In 1881 he returned to the United States and later toured with Buffalo Bill Cody. He was eventually killed in the badlands of South Dakota on December 15, 1890, by Indian policemen who had gone to arrest him for leaving the Standing Rock Reservation. Two weeks later the army killed more than 170 Sioux, including women and children, at Wounded Knee.

The buffalo that the Sioux and other Plains Indians had depended on were gone as well. Most, of course, were killed by hide and meat hunters as the railroads crossed the prairie. This commercial killing had political support from some western generals and others who calculated, correctly, that without the buffalo, the Plains Indians could no longer resist. Plus it was a moneymaker.

Some of the last few bison in the badlands of what was to be Garfield County were killed for science. In 1886 William T. Hornaday was in the Hell Creek area on a mission to kill enough bison for a display at the Smithsonian before the animal completely went extinct and there were no specimens left. Hornaday was a reluctant hunter, motivated by science, and worked hard to save the buffalo from ultimate extinction. But he killed the requisite buffalo and the exhibit, when it opened in Washington, D.C., was very popular.

Other bison during the time they were flirting with extinction were killed neither for food, hides, nor science, but for sport. Hunters who shot the bison for pleasure and trophies decried the disgraceful market hunters at the same time that they rushed to kill a bison, before there were none left to kill.

Bison never went completely extinct, although they came close. The bison alive now represent another attempt to reconstruct the past. It is more easily done than the re-creation of dinosaurs, but still, there have been glitches along the way. There are a few pure herds that represent the DNA of the bison that covered the plains in 1491. But many other herds represent animals that have been interbred with cattle in the attempt to bring them back. Are they true bison? They are certainly good to eat, but for scientific studies, and for the emotional satisfaction of knowing that the animal you see before you is of the same blood as those that native Americans ran off a buffalo jump for thousands of years, you want the pure descendants.

With the buffalo gone and the Indians defeated eastern Montana entered a brief and uncharacteristic boom period, according to K. Ross Toole, in *Montana: An Uncommon Land*. From 1880 to 1886 cattle herds fattened on the free range in relatively mild years. The winter of 1886–87 was different. Toole writes that cattle outfits suffered catastrophic losses in this "hard winter," some losing 75 to 80 percent of their herds. "January was bitterly cold," he writes. "The hope was that February would see a thaw."

But "temperatures at Glendive," which is just east of Garfield County, "from February 1 to February 12 averaged -27.5 degrees Fahrenheit." In March, after hot "Chinook" winds, "in every gully, every arroyo, along the streambeds, and dotting the level plains were the rotting carcasses of thousands upon thousands of cattle."

The open range disappeared with the coming of ranchers, who practiced cattle and sheep ranching of the sort now common here. Garfield County recovered in the 1890s and began to take on some of its current character, except, of course, that the Missouri had not been dammed, and Fort Peck Lake created, nor had the Charles M. Russell National Wildlife Refuge, generally referred to as the CMR, been established.

THE DIGGING BEGINS

The first European-American fossil collectors arrived in the West in the middle of the nineteenth century. One early collector in Montana was Ferdinand Hayden. He found a duckbill tooth in the Judith River Formation, from the late Cretaceous, in 1854. This was during the Indian wars, and according to one historical account of the early paleontologists, he acquired an Indian name that suggested the hazards of paleontology at the time, "the man who picks up stones while running."

Fossils were, however, well known to American Indians, so if he did have that name, the stones he was picking up must not have been obviously old bones. Dinosaur skeletons, and mammal and reptile and other skeletons, had been weathering out during the thirteen thousand years when the first Americans occupied the continent. And these early Americans had, of course, found the fossils and come up with their own interpretations, weaving the bones of mastodons, mosasaurs, and pterosaurs into legends of thunder beings and water monsters.

Based on the discoveries of shells and other remnants of marine creatures in places that the inland sea had covered, many Indian tribes believed that the land they were on had been underwater at some point. Like paleontologists after them, they, too, had a reconstructed past in mind.

One of the first collectors in the area of Garfield County was Barnum Brown of the American Museum of Natural History, the man who identified and named the Hell Creek Formation. He arrived shortly after Arthur Jordan, a remarkably enterprising man, who had emigrated from Scotland as a boy, and founded the town of Jordan, which began as a post office in 1899.

In 1902 Brown was sent to explore the Hell Creek area by Henry Fairfield Osborn, the paleontologist who would become president of the museum in 1908 and preside over the glory days of the museum's fossil collecting. This commission meant that Brown was well on his way to becoming one of the most successful and best-known collectors of dinosaur fossils. He had already been fossil hunting in Wyoming and had been told by Hornaday of fossils weathering out in the badlands near Jordan.

On his first trip there he found fossils of an unknown dinosaur that Osborn christened *Tyrannosaurus*. In 1908 he found a more complete specimen here that included a well-preserved skull.

The Hell Creek Formation has continued to attract fossil hunters of all sorts, academic and professional, and over the years has produced more than its share of tyrannosaurs and other dinosaurs. It has been studied in three states by paleontologists from all parts of the country. The reasons are those that I have already described, the exposed and weathered badlands. The worse the country, the more tortured it is by water and wind, the more broken and carved, the more it attracts fossil hunters, who depend on the planet to open itself to us. We can only scratch away at what natural forces have brought to the surface.

So, like many others before us, our team from the Museum of the Rockies attacked the Hell Creek rocks. Although the formation is known for being rich in *T. rex* fossils, that was not what attracted us initially, although it certainly did pay off. We chose Hell Creek because it is not only rich in fossils but richly varied in the kinds of fossils it yields. Other sites have lots of one thing, like many duck-billed dinosaurs. Hell Creek has a wide range of dinosaurs, other reptiles, mammals, and plants.

We planned a dig that would take a snapshot of this ecosystem at one location, focusing on as narrow a time frame as possible. No biologist would suggest that a living organism can be understood in isolation. Its living conditions, its food sources, predators, and countless other factors in its environment affect how it lives and how it has evolved. A leech makes no sense unless one knows about its environment and the creatures it feeds on.

Near Choteau, in the Two Medicine Formation, where we found the first nesting grounds of dinosaurs in the late seventies, we had managed to get a remarkably detailed record of what seemed to be one nesting season so many millions of years ago.

We wouldn't be able to be quite so precise at the Hell Creek Formation, but we did hope to find fossils of many dinosaur species and many animals and turtles and plants and pollen and mollusks. I recruited a dozen colleagues, senior scientists at different institutions, like Bill Clemens at Berkeley, who studies mammals; Joe Hartman, in North Dakota, whose specialty is clams and snails; and Mark Goodwin, also at Berkeley, a fellow dinosaur paleontologist. I also found private funding for what promised to be an expensive few years. We had geologists, students, plant people, Mary Schweitzer for biochemistry—all working independently toward the same goal. In the summer we would have as many as fifty people in the field prospecting and excavating what they found. We are still cataloging and studying our finds.

Even though we were set up to look for many different fossils, and we did find a variety of species, the formation is so rich in *T. rex* that in 2000 alone we found five specimens. The one that turned out to be most intriguing for research also turned out to be the hardest to get out of the ground.

On the morning of June 28, Bob Harmon, a native Montanan who was in charge of my crew at the dig, set out prospecting. He took a boat to a satellite camp and walked about a mile and a half, looking for good sites. He stopped for lunch by a cliff. After lunch he looked up on the side of the cliff and saw what seemed to be an exposed fossil bone. He scrambled up about twenty feet to a ledge, but he couldn't reach the bone, so he made his way back down the cliff and walked to the satellite camp on the shore of the reservoir.

If it were me, I would have gone back and got myself a graduate student. But Bob didn't get a graduate student. He got a folding chair. He scrambled back up the twenty-foot cliff with the chair. On the ledge he piled up some rocks, put the folding chair on top of the rocks, climbed up on the chair, and took photographs.

He spotted two other bones. That made three, and by my rule of thumb, three different bones from what seems to be the same creature mean an animal that died and was preserved in one place. Over millions of years wind, rain, and rivers scatter most bones. Finding three together is a sure sign that more from that same animal are under the surface.

The problem was that this hint of a skeleton was at the base of a forty-foot cliff, rising up from the shelf of the twenty-foot rise Bob had climbed up. I wanted to see more, but my knees have long since resigned from that sort of climbing. I brought in Nels Peterson, an engineering student and a rock climber. He brought several other climbers.

They set up a belaying station above the cliff and lowered people down. Then they lowered small jackhammers down to the climbers to begin work, to begin what turned into years of backbreaking work. Eventually, we found both hind legs, both femurs, one tibia and a fibula and a piece of jaw, and a bunch of bones going back into that cliff. All told, we collected about 50 percent of the skeleton. It was a tyrannosaur, and as I said earlier, we called it B. rex, for Bob.

That skeleton has led us farther into the past than any other. Not in time, but in the detail and depth of our understanding.

To be sure, it is the oldest *T. rex* skeleton, at sixty-eight million years, but dinosaurs go back more than two hundred million years, the origin of life more than three billion. Many, many fossils are older, but few have been studied like B. rex.

It began with the excavation, which at the time seemed like building, or perhaps taking apart, the pyramids. We, and by we I mean they, spent three years to free the bones—three years of many graduate students and numerous jackhammers, big and small.

Once we could see the bones, the job was still far from done. The fossils had to be jacketed with plaster, and since the site was so inaccessible, the enormous plaster-jacketed loads had to be lifted out by helicopter. One jacket, including the femur, was simply too big for the helicopter, so it had to be broken in two. That small fact—that we had to break the jacket in two—is what led us to look at the tissue inside the bone.

2

IT'S A GIRL!

A PREGNANCY TEST FOR *T. REX*

By the help of Microscopes, there is nothing so small, as
to escape our inquiry; hence there is a new visible World
discovered to the understanding.

—Robert Hooke

When we broke the plaster cast of the B. rex femur in two
so that a helicopter could lift it from the site of the de-
molished cliff, we exposed extremely well-preserved tissue from the interior of a fossil that had lasted sixty-eight
million years. It was that long ago that B. rex, an ovulating
Tyrannosaurus, had moved through the lush thickets and forests of a delta fed by several winding rivers. She had hatched,
and spent sixteen to twenty years growing to maturity before
she mated.

Whether this was her first mating or not, we can't tell. Perhaps she died without offspring. Perhaps she had shepherded a

clutch of eggs to hatching before. From the point of view of the present it may seem poignant that B. rex was living near the end of the 140-million-year reign of dinosaurs on earth, as if she were one of the last of her line. But she was only near the end in the terms of geological time. There were three million years to go before the end of the Cretaceous.

She died of unknown causes, but we do know that her burial was quick because her skeleton was well preserved, most of it, including the femur, encased in the tons of rock we had to remove with jackhammers. In fact, this femur was still in its matrix of rock inside the plaster jacket. Where we broke the jacket the bone had not been coated with any protective chemical, which is the common process for fossils found exposed to the elements. We paint them with a chemical preservative so that they will not disintegrate further, at least in external form and shape. But preserving the bone from further damage from water and weather may damage it for laboratory analysis, because the preservative can seep in and alter the very chemicals we are looking for.

Like so much in science, there was a bit of luck involved. Bad luck for the crew that had to break the cast open, and good luck for Mary Schweitzer, the beneficiary. I am fairly willing to break open fossils or cut thin sections to view under a microscope. I'm in favor of pulverizing some fossil material for chemical analysis. But without this unplanned break I doubt that we would have taken the B. rex femur back to the museum and snapped it in two. B. rex was a superb and hard-won fossil skeleton. Mary was looking for well-preserved fossil bone that had not been chemically treated, and she and I both had hopes for what she might find. But I'm not sure I would have picked this particular femur.

But necessity can be the mother of research material as well as invention. And when we saw the inside of the femur, and smelled it—fossils from Hell Creek tend to have a strong odor, which may have something to do with the organic material preserved—it was clear that this was prime material for Mary.

So we packed the bits of *T. rex* thighbone up and Mary took them with her to North Carolina State University, where she was starting her first semester as an assistant professor. For the previous ten years she had been studying and working at the museum, digging deep into the microscopic structure of fossilized bone tissue, and now she was leaving just about the time we were returning from the field season in August.

Mary snapped up the fragments. "I packed up the box," she said, "and brought it with me to Raleigh, and as soon as we got there my technician, Jen [Jennifer Wittmeyer]—I could not have done any of this without her—she said, 'What do you want to do first?' I said I had plans for the *T. rex* bone. So we pulled out the first piece of bone from the box and I said, 'My gosh, it's a girl and its pregnant.'

"I picked it up and I turned it over and the inside surface was coated with medullary bone. It's a reproductive tissue that's only found in birds. Birds are constrained by the fact that they have very thin bones, which are an adaptation for flight, and they make calcified eggshells," she said. There is not a whole lot of calcium available from the skeletal bones because they are lightweight, but birds need calcium for eggshells. "So," she said, "they developed a reproductive tissue that is laid down with the first spike of estrogen that triggers ovulation."

It was easy to spot, since it looked very different from other types of bone. Medullary bone is produced rapidly, has lots of blood vessels, and has a kind of spongy, porous look and feel to it. Since birds are dinosaurs, and *T. rex* is in the family of nondinosaurs from which birds claim descent, the presence of medullary bone made sense. Paleontologists had hoped to find medullary bone in dinosaur fossils, but they had not yet. If she was right in her snap judgment, this was not only scientifically important but a treat for all of us who love dinosaurs—a girl tyrannosaur.

THE SECOND EXCAVATION

And that is how the second excavation of B. rex began. The first, the old-fashioned kind, was to dig into the rock to free the fossil bone. The second excavation, of a sort that will mark a sea change in paleontology as it becomes more common, was to dig into the fossil itself, not with dental pick and toothbrush, but with the tools of chemical and physical analysis. Most of our current knowledge of dinosaurs and other extinct animals consists of the fruits of first excavations. I am not undervaluing this knowledge. In fact, it is almost impossible to overstate its value.

The work of traditional paleontology has produced a record of evolution on earth. The great skeletons that tower over museum exhibition halls are flashy, but they are mere points of data in the grand accumulation of knowledge. Fossils that show how jaws evolved or when a toe moved, or an opening in a skull appeared, are equally as important in mapping not just the existence of the past, but the process of evolution, and eventually the laws that govern its progress.

But there are now new means of tracing the past and some paleontologists are using them, although they don't seem to spread as fast as they might. As long ago as 1956 Philip Abelson reported amino acids in fossils more than a million years old. In the 1960s and 1970s other scientists pushed for the importance of molecular biology for scientists who study the past. Bruce Runnegar of UCLA summed up a new view at a 1985 conference when he said, "I like to take the catholic view that paleontology deals with the history of biosphere and that paleontologists should use all available sources of information to understand the evolution of life and its effect on the planet. Viewed in this way the current advances being made in the field of molecular biology are as important to present-day paleontology as studies of comparative anatomy were to Owen and Cuvier."

Change does not come easy, however. Scientific disciplines are more like barges than speedboats, slow to turn in a new direction. This is as true for scientists who study dinosaurs as for any others. And there are significant obstacles to moving in a new direction. For one thing, dinosaur fossils are so old that recovering biological materials from them has been a major challenge.

Of course we still excavate bones, and we need to. But we also need to look deep into the bones, into their chemistry. A first step is to narrow and deepen our vision, looking at microscopic evidence like the internal structure of bone, and moving even deeper to seek fossil molecules. Mary is a pioneer in this research, and as an inveterate digger myself, I like to think of her work in a similar framework. She is digging, too, but for her the fossil bone is the equivalent of the siltstone of the Hell Creek Formation, and the fossils she is trying to extract are not femurs and skulls but tissues, cells, and molecules, starting with protein and perhaps, one day, even moving on to DNA.

Mary had been working on the edge of this frontier of paleontological research for a good ten years by the time she picked up the piece of B. rex femur and declared the dinosaur to be female and pregnant. The path she had taken to scientific research was not a straight line from college to graduate school. In 1989, when she first audited a class I was giving at Montana State, she had just finished a science education certification program. She was married, raising three children, and working as a substitute teacher.

"I finished my teaching certification in the middle of the year. I loved going to school and I saw that Jack was teaching a course and I told him, 'I really want to sit in on your class.'"

So she signed up for a course on evolution. From her point of view the experience was mixed. "I ended up working incredibly hard, for no academic credit," she says, "and I got a C, which I still don't think was a fair grade. But it got me hooked. It really did. I realized that there was far more evidence for dinosaur-bird linkages, for evolution, for all these different things, than a layperson would begin to understand. And when

I really got to looking at that, it sort of changed my way of thinking, my worldview."

She had come to class as a young earth creationist, meaning that she believed the earth had been created some thousands of years ago. It was a view she held more or less by default. Many of her friends were young earth creationists, and although she was well versed in basic biology and other sciences, she had not studied evolutionary biology or given the subject a great deal of thought. "Like many hard-core young earth creationists," she says, "I didn't understand the evidence. When I realized the strength of the data, the evidence, I had to rethink things."

Whenever people talk about the conflict between science and religion I think of Mary. She is a person of strong religious faith that she says has only gotten stronger as she has learned more about science. Her faith is personal, and it is not something she brings up in conversation, but when asked, she is open and clear about it. She says the strength of the evidence for the process of evolution and the several-billion-year-old age of the earth is a separate matter from moral values or belief in God. She came to the study of paleontology from a background in which the assumption was that "people study evolution trying to find a way around God and his laws." Instead, she came to see science as a strictly defined process for gathering and evaluating evidence. "When I talk to Christian groups or when I teach in my class, I explain that 'science is like football.' There is a set of rules and everybody follows the same rules. The young earth creationists play basketball on the same field. It's not pretty." The essential question is whether a conclusion or hypothesis is supported by data or not. And that is separate, she says, from "things that I know to be true" in other realms, such as faith and morality.

Her approach fits well with the way I try to teach science, whether to graduate students or undergraduates who are majoring in art history. I don't present a worldview or a set of answers, but a process, a method. A discussion about the age of the earth, for example, would not begin with the answers, but with the question of how we pursue an answer, and the simple set of rules that govern scientific research in pursuit of answers. No student in a class of mine has to believe anything I say, or anything that anyone else says. But if we are doing science, we have to deal with evidence.

After Mary finished that first course, she started working as a volunteer in our lab at the Museum of the Rockies. She became more and more interested in some of the work. "I had so many questions," she says. After about a year and a half of preparing fossil material and peppering everyone in the lab with questions, it was clear that the level of her interest in dinosaurs and paleontology would never be satisfied by volunteering. Finally I said, "Mary, go to grad school. Figure it out for yourself. Stop bugging everybody about it." And she did.

Within four years she had a Ph.D., even though she was working, teaching, and raising her children. And her dissertation was the first, but not the last, time she stirred up some dust in the stuffy attic of dinosaur science.

The subject of the research, indeed the field she chose to specialize in, was a matter of chance and necessity. She turned to the fine structure of bone because it was something she could do without leaving home and children for the two months or so a full field season would require. The choice was a good one. Within paleontology the study of ancient, fossilized bone at a microscopic level—paleohistology—was a field with a great deal of promise. The potential was there for discoveries of much greater significance than the discovery of a new *Triceratops* skeleton, or even a new species, which was what she might have expected in the field.

For most of the last century or so, as the great dinosaur skeletons were uncovered in the American West, China, and around the world, paleontology has been a collector's game. The romance was in finding the new species and putting them on display for the public. Even now, a new discovery of the biggest or smallest or newest kind of dinosaur is sure to make the news.

This is not to denigrate collecting. It is the basis of the entire science of paleontology. It is how we find the past. And the collected fossils have been used in many, many ways, most importantly of all to track the course of evolution over millions of years. As we conduct vertical explorations into deep time, we find which dinosaurs came first and which later. We see how the characteristics of one kind of animal appear in later eras in descendants that branch out with new traits—what are called derived characteristics.

Thus, 160 million years of dinosaur evolution have been charted in the crest on a humerus, the tilt of a pelvis, the length of hind limbs, as well as the shape of skulls and teeth, the digits on a foot or hand, domed skulls, and weaponlike tails. They were measured and inspected, divided into Ornithischians and Saurischians and their subgroups. In the fall of 2006 Peter

Dodson, a paleontologist at the University of Pennsylvania, and Steve Wang, a statistician at Swarthmore, counted 527 known genera of dinosaurs and calculated that this represented about 30 percent of the number of genera that actually lived. That's nonavian dinosaurs.

Many of those genera, they suggested, would never be found because they weren't preserved as fossils. The fossil record, after all, is a sampling of the kinds of creatures that lived in the past. Becoming a fossil is no small trick. The organism has to die in an environment where it is buried fairly quickly, and the burial must last. Sediment must enclose the fossil and be turned into rock by time and pressure. The rock has to survive geological processes that could transform it and destroy the fossils within. And if the fossil is to be found and studied, the slow action of the earth must bring the rock and its enclosed treasure to the surface, where the elements can unwrap the gift for someone like me to find before those same elements destroy the fossil.

Fossils have always been rare and precious. And only recently has it become a common practice to cut them up or smash them to bits for microscopic and chemical study. In the early 1980s I went to Paris to learn how to make thin, polished wafers of fossilized bone that would allow a microscopic investigation of the interior structure. I was not engineering a vacation for myself. I was not a gourmet with a yearning to sample the work of great French chefs. As for travel, I would have probably chosen some desolate, eroding, fossil-rich locale in Mongolia if I had my pick of destination. Then, as now, dinosaurs were my work, hobby, and obsession. I would have been happy to learn how to make and study thin sections if I had found someone closer to work with. But paleohistology was an exceedingly small field and Armand de Ricqlès, at the Sorbonne, was my best chance as a teacher and mentor.

INSIDE THE BONES

Paleohistology, essentially the study of ancient tissues, in my case the investigation of the microstructure of dinosaur bone, had picked up speed in the 1980s, when scientists came to see many dinosaurs as warm-blooded. One of the most crucial arguments involved structures called Haversian canals, small tunnels for blood vessels. Some dinosaur bone was riddled with them, meaning that it had the kind of rich blood source that characterizes fast-growing bone in birds and mammals. Cold-blooded reptiles grow differently, and their bone looks different. Dinosaurs were beginning to look much more like ostriches than alligators.

Other findings were also important in building the case that many dinosaurs were warm-blooded, unlike other reptilians. Population structures, such as the ratio of predators to prey, and parental behavior both suggested dinosaurs were more like ground-nesting birds than any living reptiles.

By the time Mary was doing her master's work in the early nineties, we were using new techniques. CT scans of fossils showed us interior structure without doing damage to a fossil. Scanning electron microscopes let us see the smallest details. She was learning and using those techniques and more, and dinosaur paleontology had changed enough that her work did not need to take her to Paris. She collaborated with colleagues outside of paleontology in Montana and elsewhere. And of course, her techniques took advantage of the explosion in computing power that has changed all aspects of science profoundly. It is something of a shock to remember that in the early eighties, e-mail was unknown to most of us, personal computers were just beginning to become popular, and the World Wide Web was nowhere to be seen. We didn't have cell phones in Paris. In the summers, doing fieldwork, we had no phones. We relied on the ancient technology of walkie-talkies.

For her dissertation Mary wanted to study load-bearing bones in some of the large two-legged dinosaurs. From work on a *T. rex* specimen found in 1990 she concluded that the tissue in load-bearing fossil bones would be different than that of bone that did not bear weight. She wanted to test her hypothesis. What led her to go in a different direction was a happy accident, although it didn't exactly seem like that to her at first.

In order to study these bones, she was making thin cross-sections for study under a microscope. But bone, even modern bone, is not easy to work with. And fossilized bone, part rock, part preserved bone, part who knows what, was really difficult. So she was having some trouble getting the sections right.

"I had a friend in the vet lab, a bone histologist who was helping me with a problem I was having making thin sections." The friend went to a veterinary conference to give a talk on her studies of bone histology in modern animals during the time she and Mary were working on dinosaur thin sections. Among the sections mounted on microscope slides that she took to project on a screen was one of the *T. rex*, Museum of the Rockies specimen 555, or MOR 555. During the question-and-answer session she was asked what the oldest bone was that she had worked with. Funny you should ask, she said, and showed the slide of the *T. rex* femur.

Then, after the session was over, someone in the audience came up to the podium and said, "Do you realize you've got red blood cells in that dinosaur bone?"

The result, Mary said, was that Gail "called me up as soon as she got back. And she had me come over and look at it and I thought, There is no way in God's green earth that anybody's going to believe that these are blood cells." But that's what they looked like, and there they were, right in the Haversian canals where they ought to have been.

This was, in a certain sense, an inconvenient discovery, if indeed it was a discovery. Any claim for the discovery of fossil red blood cells that were sixty-plus million years old would be controversial. And Mary, whose ambition was to do a manageable chunk of research to get her master's, would have to try to prove or disprove the discovery and then defend her findings in a very public way. She did not feel ready for this kind of attention. It was a bit like being called up from the minor leagues to pitch in Yankee Stadium when you weren't sure you had control of your curveball yet.

She had a good, safe dissertation project ready to begin, solid work, but nothing that would suddenly push her into the limelight. The last thing she wanted, the last thing many graduate students would want, was to research a highly controversial claim for a dissertation. If an established scientist were to report remnants of red blood cells in dinosaur bones, that would be hard enough to defend. A dissertation that reported such an extraordinary find would inevitably draw some negative attention. Consequently, Mary held off on telling me about the apparent red blood cell remnants. She wanted to proceed slowly and carefully and to have all her ducks in a row before she approached me to discuss the apparent find. Another grad student who had seen the tissue sample told me what was going on, and I called Mary in to talk.

As Mary remembers it, I was furious. She felt she was being called on the carpet to explain this highly suspect "discovery." Mary laid out quite clearly what evidence she had. She didn't think she had any proof that these were red blood cells. But there was a lot of evidence that pointed in that direction. After a long conversation I suggested that she do her dissertation by setting up the hypothesis that these were fossilized red blood cells, and then try to knock it down.

The vast majority of vertebrate fossils found by paleontologists have been of the hardest parts of animals—bone, teeth, horn. Impressions in rock have been found of muscle, skin, and internal organs. The preservation has sometimes been remarkable.

One example was reported in 1998 in *Nature*. This came after Mary's work on the apparent red blood cells in MOR 555, but it is worth noting because of how remarkable the preservation was. This was a small theropod dinosaur, found in lower Cretaceous sediments in southern Italy. It was, in fact, reported in 1993 as the first dinosaur ever found in Italy.

The skeleton, classified as *Scipionyx samniticus,* is less than ten inches long from the tip of its nose to the end of its tail, or in scientific language, "from the tip of premaxilla to the last (ninth) preserved caudal vertebra." The skeleton seems almost complete and the internal organs, in particular the intestines, are completely visible. Cristiano Dal Sasso and Marco Signore, who described the intestine and what may have been a liver in their 1998 report in *Nature,* concluded that periods of low oxygen in a limestone deposit in a lagoon resulted in the preservation of the internal organs.

The image of this fossil, in the dry environment of a scientific journal, is still quite moving, perhaps because of its small size, or perhaps because of how complete and exquisitely detailed the preservation is. Sometimes it is tempting, given the traditional museum reconstructions, to think of dinosaurs as skeletons, not full-fleshed creatures. But this fossil skeleton, of a creature the size of a small lizard but so clearly a dinosaur, with its internal organs still visible, is a vivid, whole creature. Not living, certainly, but so fully present that it is hard to grasp how many tens of millions of years it had been in limestone before it was recovered.

Of course, there are many kinds of fossils. There are impressions in rock of plants, which, are, of course, soft. And there are microfossils, traces of microscopic life in rock. There are claims of fossil evidence of life dating back to three and a half billion years ago, although they are not completely accepted. These are impressions in rock, not the tissue of the cells.

And there are coprolites. One intriguing study was led by Karen Chin, of the University of Colorado. This also came after Mary's work on the red blood cells, and in fact Mary did some of the identification of the fossil tissue. Karen was working on a coprolite—a chunk of fossilized dung—apparently from a tyrannosaur. Coprolites are not common. And this one was unusual because it appeared to contain undigested muscle tissue from whatever the carnivorous dinosaur had been eating.

That was a true rarity. In the introduction to her paper on the coprolites, Chin referred to just over a dozen examples of fossils of muscle or skin or other tissue preserved well enough that its microscopic structure could be clearly seen. As for muscle tissue in coprolites, before Chin's discovery only two early papers, in 1903 and 1935, had reported such a find, without photographs to document it.

The coprolite was found in rock dating to the Cretaceous in Alberta, Canada. It was lying on the surface, and had been for some time, in its fossilized, rock form, since some lichen had begun to grow on it. But it was immediately recognizable. It was about two feet long and six inches wide—about a gallon and a half of dinosaur dung. And judging from an uneven surface on the underside, it appeared to have been "deposited in viscous state on uneven terrain." Usually, the language of scientific papers is so abstract that you have no idea what the authors are talking about. This description was a vivid exception.

As I said, this work came after Mary's report on the fossilized red-blood-cell remnants. They were not part of the context in which she was working, but they suggest the predominance of bone fossils, and the excitement about finding something else. Before her work, apparent fossilized red blood cells had also been reported, but rarely. One find was in two-thousand-year-old human bone. In 1939 evidence of red blood cell remnants was reported in a lizard tens of millions of years old. If anyone had found remnants of red blood cells in a dinosaur bone, I didn't know about it then, and still don't. I should point out, also, that calling the fossils red blood cells is a shorthand way of speaking that may be a bit misleading, just like calling a fossilized dinosaur femur a bone. It is not a bone in the sense that a femur of a recently dead cow is a bone. Minerals in the dinosaur bones have been replaced, chemical changes have occurred. What Mary was seeing, would, if they were real, be remnants of red blood cells with some of the original chemicals remaining, and some of the structure, but with other parts changed forever. Unlike sea monkeys, they could not be reconstituted by adding water.

This does not lessen how extraordinary it is to find any preservation of red blood cells. There are, after all, obvious reasons why, in animals, the hard parts are the ones that survive. Flesh rots. It is eaten by animals large and small. Any creature left on the surface is quickly dressed down to a skeleton. Even when a dead animal is buried, insects, worms, microbes, and chemical disintegration usually leave nothing but bone. Despite occasional reports of cell-like structures in some fossils, the idea of doing a dissertation announcing the finding of red blood cells was shocking.

In the paper based on her dissertation, which Mary published with me as one of the coauthors (a common role for dissertation advisors or supervising scientists), Mary eventually claimed only that there were heme compounds in the fossils, parts of hemoglobin molecules, indicating that the full hemoglobin molecules had been, or were still, in the bone. Hemoglobin is a protein that enables red blood cells to bring oxygen to muscles. More hemoglobin is what bicyclists competing in the Tour de France are after when they use illegal so-called blood doping techniques.

Part of the reason for concentrating on hemoglobin and byproducts is that the presence of chemicals can be tested with established procedures. It would support the idea that what we were seeing were remnants of red blood cells, but did not require determining how much of cell structure had survived and what the degree of fossilization was. Fragments of hemoglobin molecules had been found before in bone a few thousand years old, and blood residues on stone tools up to a hundred thousand years old.

CHEMICAL TRACES

There were other discoveries that made it seem reasonable to look for preserved protein molecules, like hemoglobin, in truly old bone. Since the 1970s sequences of amino acids—which are strung together to make a protein molecule—had been found in mollusk shells that were 80 million years old, and in dinosaur fossils from 150 million years ago. Recent work on the biochemistry of fossils had led to discoveries of a bone protein, osteocalcin, in dinosaur bone.

It was necessary to marshal a variety of techniques that had seen little use in paleontology, such as liquid chromatography and nuclear magnetic resonance, to test for hemoglobin. After finding the chemical and physical signatures of parts of hemoglobin in the bone, but not in the sandstone surrounding it, Mary sought biological evidence and sent another lab extracts of fossil bone material. Rats injected with the extract made antibodies—against avian hemoglobin. That is to say, their immune systems recognized something foreign and brought forth weapons—antibodies—specifically tailored to meet and disable the new invader. Using antibodies for testing is a common practice, so it was possible to determine that these antibodies would also work against hemoglobin from birds—not mammals or reptiles, but birds. This was completely consistent with the dinosaur bone's containing elements of hemoglobin from dinosaurs, and ruled out contamination by humans or other mammals either in the lab or before the bone was collected.

So it seemed that hemoglobin or products from the breakdown of hemoglobin were still present after sixty-eight million years.

The evidence pointed to ancient molecules from red blood cells that had been preserved. Were the structures that she had observed fossilized red blood cells? This was the best explanation we could arrive at. But it was presented as a tentative conclusion, open to any challenge that other scientists could think of. Essentially, there was a lot of evidence that was consistent with the idea that the structures observed under the microscope were red blood cells, but not enough for a definitive assertion that these structures were fossilized cells.

"I couldn't disprove it," Mary said. "I couldn't prove it. At this point in time I still don't know what those things are, and might never know. But the research did get me into the mind-set of thinking of fossils as something other than fossils. I don't treat fossils like fossils; I treat them as I would modern bone."

The reaction was quite strong. Unfortunately, the biggest outpouring of interest was from creationists. They absolutely loved the idea that the bones had some remnant of red blood cells.

They argued that since we had thought that such things couldn't be preserved and now had found they were preserved to some extent, that meant our dating was wrong. They ignored the accumulated evidence of geology, radiometric dating, and numerous other facts that made clear that what we were wrong about was not how old the fossils were, but the possibilities for preserving soft tissue.

In retrospect, Mary, who took the brunt of the attack, points out that science was in part to blame for its acceptance of the conventional wisdom that no biological materials like hemoglobin or red blood cells could survive as fossils. We didn't know what we thought we knew. That's common enough in science, which is not a collection of answers but a process of posing questions and then coming up with more questions. Knowledge is always provisional. It is not that previous answers are overturned so much as that they prove to be incomplete or not so widely applicable as they might have seemed.

The knowledge about how flesh and bone disintegrate was, indeed, gathered by observation and experiment. We all know what happens to an animal body left out in a field or on the road. We can see dried-out skulls in the desert, crumbling deer bone in the forest. Everything dies and everything falls apart and decomposes. Scientists have tried to quantify this process by setting out the body of a dead animal and carefully tracking its decomposition.

"We've got body farms," says Schweitzer, "where we know how tissues degrade. We know how long it takes under different environmental conditions." Laboratory research has tracked the same processes at the cellular level. "We know how long it takes for membranes to break down. We know how long it takes for the nucleus to go away. We even know the cellular kinetics. We know the enzymes involved. We know how they interact with one another. We know how cells degrade. We know how proteins degrade. We know how tissues degrade. We know it. Well, they know it, I'm not that smart."

But these studies are all done on muscle, skin, and other soft organs. Not bone. "Nobody in their right mind works with bone because bone sucks. It's really hard to work with." So, says Schweitzer, the models of how things fall apart, from the large scale to the small, are not based on bone. "They are not based on the microenvironments inside bone. And when you put a cell or a tissue inside a mineral, you change everything. You change the ability of enzymes to attack, you change the ability of microbes to get in and eat. . . . And nobody looks at it because bone is the pits to work with. But since bone is all we have from dinosaurs, I look."

The gender question would be the research Mary and Jen undertook first. What Mary saw was a specific kind of bone that is known to be created in birds as they are producing calcium-rich eggshell. Because of the close relationship between birds and dinosaurs, paleontologists had predicted that this tissue would be found in dinosaurs as well, but no one had yet seen it. Mary and Jen compared the B. rex bone to ostrich bone, again using the scanning electron microscope as well as the light microscope. The bone was clearly the same and the conclusion was clear.

The microstructure of the fossil was the same as that of medullary bone, which is very rapidly deposited. "It has tons and tons of blood vessels," Schweitzer said. "One of the things that's standard for modern bone studies, when you want to get at the architecture, the microstructure of the bone at the level of the protein, you want to remove the mineral. So I told Jen, 'We want to etch the bone, but don't leave it in there very long.' Like everyone else, I thought that if you take away all the mineral from dinosaur bone, you would have nothing left, because of course the organic molecules don't preserve." Etching means to put it in an acid bath. So the bone only needed to take a brief dip to clean it up.

SPROING!

"When she went to stop the etch, she went to pick up the bone and put it in the water, it went *sproing!*" This was not a large piece of bone. Jen was picking a small piece out of the acid bath with very fine tweezers under a dissecting microscope. Schweitzer went to see for herself.

"It bent, it twisted, it folded," she said. "It was the most bizarre thing I'd ever seen. I said, 'I don't believe this is happening—do it again, please.' She did it on the second piece and it went *sproing*." What the material seemed like was collagen, but to identify a specific protein like collagen required gas chromatography, mass spectrometry, and biological assays as well. That was a research project in its own right. So she set that suspicion aside for the moment and looked to see what she could find in another piece of bone, not the reproductive tissue that made B. rex unique, but the cortical bone all dinosaurs and all four-limbed vertebrates have. Jen set up demineralization baths for the new samples and checked their progress.

Under the dissecting microscope the demineralized material, washed in distilled water, had what looked like fragments of tubing, very, very small tubing. And if she picked up a piece of this material she could see it move back and forth under the microscope. It was flexible, some kind of flexible, transparent tubing from sixty-eight million years ago.

Jen came back to tell Mary what was going on. "She said to me, 'You're not going to believe this, but I think we have blood vessels.' I said, 'You're right. I don't believe this. Nobody's going to believe this. We can't talk about this.' I don't think either one of us slept for three weeks. We kept repeating and repeating and repeating our experiments. We actually hardly talked to each other."

Further demineralization and washing showed that the flexible vessels were transparent. Jen continued working to compare the *T. rex* material with demineralized ostrich bone. Ostriches and emus are among the most primitive of living birds and presumably marginally more closely related to non-avian dinosaurs. The ostrich bone yielded the same kind of vessels. Furthermore, both the dinosaur and ostrich vessels contained small round red structures. And in the dinosaur, some of those round structures had dark centers that looked a lot like cell nuclei.

Using a scanning electron microscope to get a closer view, similarities between dinosaur and ostrich bone were just as strong. And between fibers in the bone matrix of the dinosaur, Jen and Mary found something even more surprising, the unmistakable outline of the cells that make bones grow—osteocytes.

These cells secrete the minerals that make up the hard scaffolding, or matrix, that makes bone rigid and strong. As builders, however, they strand themselves inside the structure they are producing. How, then, to get the energy needed to keep working and to get rid of waste material? The cells have unmistakable tendrils called filipodia that extend far out from the central body of each osteocyte and connect to filipodia of other osteocytes. They are strung out in a network that transports nutrients in and waste out. "It's like a bucket brigade," says Mary.

When Mary was first working on this material, she called me up to say she had found osteocytes. I assumed she meant the spaces where the osteocytes would have been, which is what I suggested.

"No, Jack, actually we have the cells and they have filipodia and they have nuclei."

"Mary, the freaking creationists are just going to love you."

"Jack, it's your dinosaur."

To continue to check these results Mary used the same methods of demineralization and light and electron microscopy on two other tyrannosaurs and a hadrosaur, or duckbilled dinosaur. She and Jen found vessels and structures that looked like osteocytes in all of them. But not all the vessels were flexible and transparent. Some of them were hard, crystalline shapes and some were just the same as the vessels in B. rex. It was hard to imagine that these microscopic remnants of vessels could be preserved, perhaps as some kind of flexible fossil with chemical substitutions for the original material of the vessel. But it was even harder to imagine what else the apparent vessels might be.

The findings on the presence of the transparent vessels and the gender of B. rex were published in two papers in *Science* in 2005, March 25 and June 3, just two months apart. Mary, her technician, and I were among the coauthors. One question that was not answered in these papers was exactly how you could end up with a blood vessel still flexible after so many millions of years. How was it preserved? We don't know. Research to answer that question is going on now. Certainly, these are not untouched and unchanged blood vessels. They have been fossilized but remained flexible. Or so it seems so far.

Nothing in science is ever accepted until it has been replicated by other scientists, and so far this has not happened

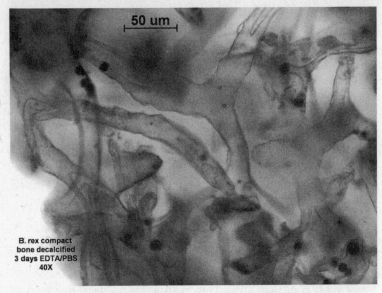

50 um

B. rex compact
bone decalcified
3 days EDTA/PBS
40X

Transparent vessels from a sixty-eight-million-year-old fossil bone from the B. rex skeleton.
The bone has been decalcified and the sample is magnified forty times. The round, dark
shapes look similar to red blood cells seen in an ostrich bone similarly treated.

with Mary's discoveries. She is the first to say that although
these may seem to be blood vessels with everything we know
now, contrary evidence may be forthcoming. In fact, she has
admitted that she doubted her findings and that consequently,
she and Jen set out to see if they could find similar materials in
other bone and fossil bone.

"We started with a chicken that we went and got at the gro-
cery store. And we worked our way back to Triassic bone."
What they found was that the flexible material that seemed
like blood vessels was "amazingly prevalent." That, she said,
"makes me really nervous." Why would it be so common? The

question is still unanswered, but alternative explanations are no more persuasive, she says.

At the 2006 meeting of the Society of Vertebrate Paleontology an independent dinosaur researcher, Tom Kay, presented a poster that argued that she had found biofilms.

"So what Tom proposes is that they're microbial biofilms. But I'm not sure I buy that, because we have looked at material that crosses taxa, that crosses continents, that crosses depositional environments, that crosses temporal ranges, and the vessels all look the same.

"There are a million reasons why this stuff is not consistent with biofilms. For one thing, Tom hasn't shown that biofilm will grow in bone, or that it will form flexible tubes. If it is modern biofilm then it is defined by the presence of cell bodies, which he didn't demonstrate. Biofilms are rather thin and grow kind of patchy because they get nutrient limited. No one has shown that they will form interconnected hollow tubes to the extent that we have observed. And, biofilms are rather homogeneous in texture, whereas we show textural differences between the osteocyte filipodia and the matrix they extend into.

"So he's challenging our hypothesis, which is great and I applaud it, but he hasn't produced any data to support his alternative hypothesis that's any more valid than mine." Further challenges have been made, but so far, neither Mary nor I see any effective criticism that undermines the idea that these are remnants of blood vessels. The next step in her research was a much larger one, and that was to tackle the apparent presence of collagen. In some ways this was more important than either of the previous findings, because of the nature of proteins, and what they can tell us about different species and their evolution. This was to take the second excavation of B. rex a level deeper, and move further in the complex discipline of molecular paleobiology.

3

MOLECULES ARE FOSSILS TOO

BIOLOGICAL SECRETS IN ANCIENT BONES

Miracles from molecules are dawning every day,
Discoveries for happiness in a fab-u-lous array!
A never-ending search is on by men who dare and plan:
Making modern miracles from molecules for man!

> —"Miracles from Molecules," theme song of *Adventure Thru Inner Space,* an attraction at Disneyland from 1967 to 1985

urassic Park was released in 1993 and was an immediate success. It grossed more than any previous movie, taking in more than $900 million worldwide. The film was an adaptation of Michael Crichton's novel of the same name, published in 1990. In the book and movie dinosaurs are cloned through the recovery of ancient dinosaur DNA. It was a classic piece of science fiction. Crichton started out with real science and stretched it beyond current boundaries, into the realm of myth and fairy tale, producing a story of universal appeal. Monsters of the unimaginable past brought back to vivid and dangerous life.

Crichton imagined that a mosquito that fed on the blood of dinosaurs was trapped in amber and preserved. Not only was the insect preserved in amber, but so was the dinosaur DNA in its gut. Scientists implanted the ancient DNA into a frog's egg. In 1993 frogs had been cloned at the tadpole stage, but no other vertebrates. So the technique made sense. In the story the dinosaur genes take over and the egg grows into a full-blown dinosaur.

In some ways the book and the movie were prescient. In 1997 a mammal was cloned for the first time, a sheep named Dolly. Since then a number of mammals have been cloned and one species of fish, a carp, but no reptiles or birds as of this writing. I don't think Crichton saw the cloning explosion coming; few scientists did. But cloning was and is a favorite science fiction topic.

In other ways the book was very much of its time, reflecting a stage of the revolution in molecular biology and computer science. The human genome project was started in 1990, at the time *Jurassic Park* was published. This was an attempt to map all of the genes that go into the making of a human being. It was an obvious project, given the state of the science. But it was breathtaking in its ambition, and the notion that one could compile the set of instructions that would form a human being was, and is, shocking. Certainly no scientist would say that genes alone make a human. Genes are always affected by environment. And there were and are long stretches of DNA with no known function. Furthermore, since then, the question of how genes are regulated has become ever more important. But the dominance of the gene was at its height then.

The first genome of a multicellular organism, the millimeter-long roundworm *C. elegans,* was completed in 1998. The fruit fly genome was decoded in 2000. A draft of the human genome was released that year, and a finished, although not really complete, genome was released in 2003. None of this could have been done without the discovery of DNA and the development of chemical techniques to take it apart and determine the sequence, and computer power to massage and understand the information gained from the biochemistry.

The ability to read stretches of DNA, to compare them to other stretches of DNA, to fish out genes and identifying markers of different species, changed the way all biology was done, and not just the biology of living animals. It also began to change the study of the history of life, both in raising the value of finding traces of biological molecules in fossil bones, and in promising a new treasure trove of information about the past in the genomes of living organisms. A genome does not tell you everything about an organism. Much DNA doesn't fall into the categories of genes as we define them, self-contained lengths of DNA that contain the code and starting and stopping instructions for making RNA that in turn is translated into protein molecules.

But a genome tells you a great amount and it made complete sense, given the state of molecular biology when he was writing, for Crichton to pick the method he did for re-creating a dinosaur. Ancient DNA seemed like a good bet, at least for a novel. And sometimes science imitates science fiction. In 1994, four years after *Jurassic Park* was published, and a year after the *Jurassic Park* movie was released, Scott Woodward at Brigham Young University reported the recovery of DNA from eighty-million–year-old fragments of dinosaur bone found in a Utah coal mine. The report was quite a surprise. Crichton put the preserved dinosaur DNA in *Jurassic Park* in the gut of a mosquito safely encased in fossilized tree resin because that was far more believable than dinosaur DNA being recovered from a fossil bone buried for tens of millions of years.

Somehow, Woodward argued, the fossils he found in the mine had been protected from biochemical reactions and other forces that would have caused the DNA to fall apart. There was some infiltration of minerals to replace biological chemicals. But when the bone was examined under a light microscope, Woodward reported, bone cells and possible cell nuclei were visible.

Taking care to preserve sterility and not to contaminate the bone with human DNA (which, as we all know from watching *CSI*, can be found on cups and cigarette butts), he and his colleagues took very small pieces of bone, turned them in powder, and then turned to what may be the most important technique in the molecular biology revolution, the polymerase chain reaction, known affectionately to scientists as PCR. The process, by which stretches of DNA can be multiplied by the hundreds of thousands, enables scientists to identify traces of DNA. A scientist named Kary Mullis won a Nobel Prize for discovering the enzyme that makes the process possible. Mullis, who lives in southern California, may be the only Nobel laureate who takes surfing at least as seriously as science. Indeed, Mullis is as unconventional as his technique is irreplaceable. He has written about his own experimentation with drugs, about possible experiences with space aliens, and has said that he does not believe the human immunodeficiency virus, HIV, causes AIDS. I'm not sure he actually believes everything he says. I know him, and I think he likes to provoke discussion. But he does say some outrageous things.

PCR relies on an enzyme Mullis discovered in bacteria in warm springs in Yellowstone National Park. The bacteria is called *Thermus aquaticus* and its enzyme is Taq polymerase. It is valuable because it comes from an organism that evolved to tolerate high temperatures. Consequently Taq (short for *Thermus aquaticus*) polymerase stays stable at high temperatures.

In preparation for initiating the polymerase chain reaction DNA is heated so that the two strands that make up each molecule separate. Molecules called primers target a stretch of DNA for which they have been designed, and serve as a marker, to tell the polymerase enzyme where to start copying. So the marked section of DNA is duplicated, and each single strand becomes a double helix again. This solution is again heated so the two helixes will separate into four strands, which the polymerase turns into four double helixes and the next heating turns into eight single strands.

In undergoing the heating and cooling steps over and over again, the target stretch of DNA keeps being duplicated, the total growing in a geometric progression—2, 4, 8, 16, 32, 64,

128. In a couple of hours more than a million copies can be made of a target gene. Making so many copies provides enough DNA to be sequenced, that is, to have the code read. Computer power allows the information gathered through biochemical reactions to be processed to determine the sequence of nucleotides in the gene.

Woodward used this process to amplify DNA from his fossil bone samples. He concluded that the DNA had not come from modern birds, mammals, or reptiles. Nor, he claimed, were they the result of bacterial or human contamination. He concluded that he had likely recovered DNA from Cretaceous era dinosaur bones. Or, as he put it in the cautious conclusion to his scientific report, "The recovery of DNA from well-preserved Cretaceous period bone may be possible."

The response was the scientific equivalent of bench clearing when the opposing pitcher plunks someone on your team. Several critiques were published by the same journal that had published his initial report. The authors had reanalyzed the data that Woodward had presented, in particular the claims that the fragments of DNA were not mammalian. Mary wrote one of the responses along with Blair Hedges at Penn State, and other responses came from researchers in the United States, Germany, and the Netherlands. All the critics concluded, and they convinced me, that the samples of DNA Woodward had amplified could well be mammalian, and if any mammalian DNA were present, that would mean contamination, either ancient or contemporary. By far the most plausible explanation for his results, the critics argued, was that the handling of the samples had allowed contamination with human DNA. Nobody has repeated his findings, and nobody has found DNA in fossils of that age since.

DNA has, however, been retrieved from extinct animals, such as mammoths frozen in permafrost and, arguably, from Neanderthals. But tissue frozen in ice, or at most a few tens of thousands of years old as in the case of the Neanderthals, is nothing like an eighty-million-year-old fossil. Contamination is a particularly sticky problem in attempting to sequence the DNA of Neanderthals, since they were so close to us. They belonged to the genus *Homo* and were either a different species or a subspecies, meaning that their DNA would be very hard to distinguish from that of modern humans. Consider that the chimpanzee genome is 98 percent the same as the human genome, at least. The Neanderthal genome, presumably, is much, much closer. So tracking down the ever-present possibility of contamination is extremely difficult. And with chimps, at least we have some idea of the differences. With Neanderthals we don't even know where the differences in DNA lie.

Just as this book was going to press, the sequencing of the complete mammoth genome was announced. Reconstruction of the mammoth genome was possible because of the frozen tissue that had been preserved and advances in sequencing technology. DNA from ancient sources is recovered in fragments that can be fed into new sequencing machines developed only in the last decade that allow the amplification and sequencing of shorter fragments of DNA than were useful before. These machines are expected to improve rapidly, since computer power grows ever better and cheaper. With future generations of such technology the cost of sequencing the whole genome of an individual will drop precipitously. As it is, James Watson's genome was sequenced at a cost of only $2 million, compared to the $3 billion cost to complete the Human Genome Project, which produced the first sequence of a human genome. The cost of sequencing genes is dropping rapidly. It was offered at $350,000 in 2008 and one company had two takers. Scientists and, perhaps even more eagerly, entrepreneurs are looking forward to the day when anyone can get his or her genome sequenced for $1,000.

Once the cheap, or rather, affordable genome mapping becomes available, many people will have their entire genome sequenced. No doubt most of them will be disappointed by how little can be done about what a genome tells us, and by how little it will tell. Traits like intelligence and athletic ability have not been pinned down to any set of genes that prospective parents might want to provide their children. That may be just as well. Eugenics, selecting for only particular human traits, inevitably diminishes those not selected for. The movie *Gattaca* that came out a few years ago, starring Ethan Hawke, is a nice tour through some of these issues. It presents a world where everyone is healthy, and no one has "defects," at least no one with a good job. The natural-born children, whose genes are the result of the old-fashioned parental roll of the dice when sperm meets egg, are restricted to menial jobs.

The ability to sequence genomes is a powerful tool, and its future uses are unknown. So far, however, it has challenged, but not conquered, the tyranny of time, which still ravages the biochemistry of fossils. The farther back one goes in time, the less DNA is available. A common rule of thumb had been that a hundred thousand years was a limit for recovery of DNA, with the length of DNA molecules being gradually eroded through a variety of chemical processes as bones were fossilized.

If fragments of DNA are ever recovered from fossils that have come down to us from deep time, they would be valuable even if they were only hints and snippets. A few years ago, Mary wrote an article about the future possibilities for paleontology, concentrating on pursuing molecular fossils. She wrote that "if a two-hundred-base-pair fragment of DNA (e.g., the hemoglobin gene) with forty informative sites could be recovered from exceptionally preserved bone tissues of a *Velociraptor*, it would be possible to align the dinosaur gene region with the comparable region of extant crocodiles and birds."

This could help track the course of evolution. Small sections of a gene in a dinosaur could be compared with small sections of a gene in a bird or crocodile. The pace at which genetic change occurs during evolution could be determined definitively by the amount of change in that fragment.

The DNA Woodward found may not have been dinosaur DNA, and in fact, finding truly ancient DNA, more than a million years old, may be a tantalizing but unreachable goal. But he was on the right track in the sense that many researchers have been finding that biological molecules can last longer than had been thought and open a new window on the past. They haven't found DNA from dinosaurs, but they have used new technology to find other molecules that have survived the eons.

DISCOVERING FOSSIL MOLECULES

It was only in the past half century that scientists really began looking for ancient biochemicals. Philip Abelson was one of the first. A physicist who contributed his ideas on the enrichment of uranium to the scientists working on the Manhattan Project, Abelson switched to biology after the war, entering the field of biophysics, a new branch of research. He was in the Department of Terrestrial Magnetism at the Carnegie Institution when he started this work and became director of the institution's Geophysical Laboratory. "Until recently," he wrote in a 1965 issue of *Scientific American*, "it was thought that the hard parts could tell us little or nothing about the chemistry of extinct organisms." But he reported finding "organic material in fossils as old as three hundred million years." In the 150-million-year-old fossil vertebrae of a *Stegosaurus* he found a half-dozen different amino acids. He also reported work to assess the potential longevity of different amino acids. After testing the rapid degradation of alanine at 450 degrees centigrade, he wrote, a projection of that information based on a well-known and often-used formula suggested that at room temperature alanine would survive for billions of years. He found amino acids in fossilized remains of horses, scallops, snails, dinosaurs, and fish.

That 1965 article was a call for more work in this new field of research. Progress, however, turned out to be slow. In the late 1950s and the 1960s all biology was in the shadow of the explosion of work on DNA prompted by Watson and Crick's discovery of the structure of DNA in 1953. Progress in DNA research was so rapid that by the early 1970s researchers were beginning to fiddle with the genes of existing life-forms, and the prospect of what was then called recombinant DNA, and is now referred to as genetic engineering, became science fact rather than science fiction.

Today disputes still rage over genetically modified food and over what sort of prenatal genetic selection or even modification of embryos might be ethical. Scientists saw the new age coming and decided to tackle the ethical problems posed by molecular genetics themselves rather than wait for Congress. Molecular biologists held a landmark meeting at the Asilomar Conference on Recombinant DNA in Monterey, California, in 1974 to begin to talk about the possible dangers of combining DNA from different organisms. The biologists and other professionals agreed on a variety of safety measures, such as physical containment when risk was high, and the use of laboratory organisms like bacteria that were effectively prevented from existence in the wild by being dependent on specific laboratory conditions.

With the future of humanity and the planet being discussed by those at the frontiers of biology, dinosaur scientists and other paleontologists largely stuck to rock and bone collecting, with some progress in extracting ancient chemicals. In 1974 proteins were found in seventy-million-year-old mollusk shells. Amino acids were found later in fossil bones. New techniques were developed, using the reactive nature of immune-system chemistry to identify biological materials in fossils. Collagen was identified, and albumin and other proteins. Hemoglobin was found with archaeological materials, old bones and old tools. As I described earlier, Mary reported in 1999 on the possible presence of hemoglobin in that first dinosaur bone she worked on. Even though creationists have said that she found actual red blood cells, what she and I reported in that paper was the presence of features that looked like they could be fossilized red blood cells. And chemical evidence of hemoglobin that was not definitive.

It became clear to some of us in paleontology that it was time for a change in the way we did our work. We didn't need to give up the satisfying summer fieldwork, the digging up of the past, but we did need to add new tools. And we needed to go beyond the dissecting microscope, through which we could see fine details of bone structure. We needed to get down to the level of molecules in fossils—and in living things. By the 1980s molecular biologists were already using differences in genes in living creatures to calculate rates of evolution and to date events in evolution. They had developed a new stream of evidence to compete with or supplement the fossils weathering out of the earth.

Clearly there was a vast amount of evolutionary information in the molecules, and paleontology had to adapt to the new world if it was to stay valid. Abelson made his pitch in 1965. Twenty years later, in 1985, Bruce Runnegar of UCLA made a similar call for paleontology to change in an address to a meeting of the Palaeontological Association, which is based in the United Kingdom. And after another twenty years—

actually, twenty-seven—three scientists, Kevin J. Peterson of Dartmouth, Roger E. Summons of MIT, and Philip C. Donoghue of the University of Bristol, made another call for change, citing Runnegar's earlier address.

Each time, of course, the emphasis was different. The emphasis now is heavily on the importance of connecting embryonic development with evolutionary patterns, the essence of evo-devo, which I described briefly in the introduction.

But they also wrote about the possibility of finding ancient DNA and using it to help track evolution. They limited the realistic prospects of DNA recovery to thousands of years, however, with deep time essentially unreachable because of the instability of DNA. They did acknowledge a realistic way of reaching into deep time that doesn't involve DNA: that is, to search for preserved proteins from tens of millions of years ago. In this they paid explicit tribute to the work done by Mary and her colleagues on ancient collagen.

In the last chapter, I described Mary and her lab tech, Jennifer Wittmeyer, noticing the springiness of the microscopic remnants of sixty-eight-million-year-old fossil bone left after mild acid had been used to leach out the minerals. They thought of collagen. It had been reported in ancient fossils and so seemed a possibility. With nothing left of the rock in which the fossil had been embedded and none of the minerals that had seeped into the bone itself, what was left could be bent and twisted by delicate tweezers under a dissecting microscope.

Collagen injections are now a staple of cosmetic plastic surgery. In reading advertisements and promises for the benefits of one or the other of the several collagen treatments, you can

almost hear the subliminal whisper that here is a miracle substance. Well, it may not be a miracle but collagen is certainly a
marvel. It is the most common animal protein. Twenty to
twenty-five percent of all protein in mammals is collagen. It is
a major component in bone and the main component in con-
nective tissue in vertebrates. It is, quite literally, what holds
our skeletons together.

It is hard to overstate the importance of proteins in living
animals. Collagen, for example, has a structural role. Other proteins transport nutrients, oxygen, and metabolic waste throughout the body. They also promote all sorts of biochemical
reactions, regulate growth and other processes, and are
important immune-system chemicals. Antibodies are proteins.
Proteins are made up of smaller molecules—amino acids—
which in turn are made up of atoms of carbon, nitrogen, hydrogen, oxygen. Sulfur is also part of some amino acids.

Not only are proteins so important in the metabolism of the
animal body, they offer a coded record of parts of an animal's
DNA. And we know the code—it is the one in which genes are
written. The genetic code contained in DNA is usually represented as the four letters G A T C. The letters refer to guanine,
adenine, thymine, and cytosine, chemicals called bases. Each
DNA strand is a string of bases that fits together with another
string in the famous double helix shape, because the bases link
to each other. They do so in a predictable way: G pairs with C
and A with T.

There are nearly countless possible sequences of bases that
can make up a gene because genes are of different lengths, so
the bases can appear in different sequence and numbers. These

genes, these sections of DNA with various arrangements of different numbers of bases, provide the instructions for all the many, many protein molecules that are found in living things.

The way these instructions are read is a process described in every high school biology text, thanks to the several Nobel laureates that figured the whole thing out. The double-stranded spiral shape of DNA is pulled apart and the single strand of DNA is copied by the cell machinery to produce a mirror image not of DNA, but of a similar molecule called RNA, or ribonucleic acid. In RNA another base occupies the place of thymine. It is uracil, or U. The RNA molecule can then be read by cellular machines called ribosomes that translate G, C, A, and U into twenty different amino acids in the sequence and quantity specified by the genes. In this part of the code three bases code for an amino acid.

A code that can be read forward by the cell to produce a protein can be read backward today. This usually requires some sophisticated machinery and techniques involving mass spectrometry. The makeup of a substance can be determined by looking at the proportions of nitrogen and other chemicals in the tissue. Each protein has a distinctive profile.

Protein molecules have distinctive shapes, as well, that relate to the roles they play in bodily chemistry.

COLLAGEN

Collagen is a large, very strong molecule made of up fibrils, which are bundled together like multistrand rope in tissues. The long, ropelike molecules have great tensile strength and find their use in bone, tendons, cartilage—many tissues that need strength and some flex as well.

Collagen has also become one of the prime biochemical fossils. It can last over millions of years, as had been shown before Mary began her work. It can be extracted from fossil bones with great difficulty, and then it is possible to find out the distinctive sequence of amino acids in this version of collagen. Because substances like collagen evolve over time just like the shapes of limbs and the number of toes and the length of a beak. Bird collagen is subtly different from mammal collagen. And within mammals, or birds, collagen may undergo changes as well. By identifying different sequences in different collagen molecules and analyzing how long the changes may have taken to occur, scientists can have a new benchmark for evolutionary change. Just as the shape of a tooth may mark the change from one kind of mammal to another, so changes in collagen may one day be tracked to the transition from birds to dinosaurs. Or, the similarity of bird and dinosaur collagen may help attest to the strength of this evolutionary link.

When geochemists began to turn their attention to fossils as well as rock, they found that just as ancient rock contained the fossil bones of animals, the fossil bones contained their own microscopic histories. So, a few geochemists added the prefix bio- to their discipline and began to prospect within fossils in the way that most of my colleagues and I prospect in the Hell Creek or Two Medicine Formations.

One of these biogeochemists, and a pioneer in her field, is Peggy Ostrom at Michigan State University. "Molecules are fossils too," she says, as a kind of challenge and manifesto.

"They can persist over time. They have shapes and sizes. They're beautiful, just like the bones. Indeed they have form and they have function." And they have a wealth of information to provide about the history of life.

Peggy Ostrom has targeted a different protein, a component of bone called osteocalcin. She chose it partly because it is small, with only forty-nine amino acids. It is also found only in vertebrates, so there is no worry about contamination from bacteria or fungi. At a meeting of the American Association for the Advancement of Science a few years ago, Peggy explained how she goes about looking for ancient osteocalcin in fossil bone.

Starting with about twenty milligrams of bone, she chemically extracts the biochemicals in the bone from the mineral matrix and then crystallizes it. The next step is a mass spectrometer, which can detect different proteins.

There are various kinds of mass spectrometry, and the technique, despite the fact that it is mentioned frequently on television, still has an aura of magic about it. Take a smidgen of something, pop it into a machine, and then, presto, you know who killed Colonel Mustard because of the traces of cyanide powder on the butler's gloves. The reality lacks magic, but not wonder. And although the techniques can be complicated, the essence of the process is simple. All elements, like hydrogen, oxygen, gold, and copper, have different masses because they are made up of different combinations of protons, neutrons, and electrons. Molecules made up of various combinations of atoms have different masses as well. In mass spectrometry a substance is vaporized and the gas molecules are hit with electrons, breaking them up into atoms. In the process the atoms are ionized, giving them a positive electrical charge. Then a detector sorts out the different masses and the data is fed into a computer to produce a graph that shows the makeup of the sample in terms of the different elements. Each molecule has a specific signature or chart, and even different versions, or isotopes, of the same molecule have different signatures.

Peggy Ostrom and her colleagues used mass spectrometry to investigate samples of fifty-five-thousand-year-old bison bone preserved in permafrost. They found indications of osteocalcin. Then they used enzymes to cut the protein they thought they had found into its component parts to get a "fingerprint" of the amino acids. She was able to do this to such a level of detail that she could identify a change in one amino acid in the osteocalcin found in the fossil bison and a modern cow. The fifth amino acid in the string of forty-nine was tryptophan in the bison and glycine in modern cattle.

She and her colleagues have also sequenced osteocalcin in the fossilized bone from an extinct horse found in Juniper Cave, Wyoming. The horse, *Equus scotti,* is dated to forty-two thousand years ago. Comparisons of the fossil molecule with modern horses and zebras showed that the osteocalcin molecule in horses had not changed, and comparisons between modern horses and zebras showed that no change had occurred in the molecule in the last million years.

So far, Peggy has gone back as far as half a million years to find osteocalcin in musk oxen fossils. Fossil molecules, she has said, are "beautiful, just like the bones, only the fact is, you can't see them." She continued, "These ancient proteins are windows into the past for us. We can now do genetic time travel. We can now, instead of looking at modern organisms to figure out how they're related, we can go back in time and actually look at the real molecules," thousands or hundreds of thousands of years old. What about sixty-eight million years old? That was the time Mary wanted to travel to with her bits of fossil from B. rex. Half a million years ago North America was filled with mammals, some we would find unrecognizable, but others that would seem familiar, like bison. Humans, in the taxonomic sense, which is to say, species in the genus *Homo*, had been around for at least a million years. *Homo erectus* had spread from Africa to Asia and Europe using stone tools, and perhaps using fire. A hominid described as an archaic form of *Homo sapiens* had appeared, somewhere between *Homo erectus* and anatomically modern humans, who appeared about two hundred thousand years ago.

It would be hundreds of thousands of years after the time of those fossil musk oxen before behaviorally modern humans left Africa (around fifty thousand years ago) and eventually took over the globe and started digging up fossils of its own ancestors. Still, five hundred thousand years ago, though not truly comprehensible as an expanse of time, leads us back to a moment in the life of the planet when the scene would at least have been comprehensible to us.

INTO DEEP TIME

Mary was trying to make a leap into deep time. And every difficulty in testing fossil bone increases with increasing age. What if the sample is contaminated, what if the tests are wrong, or inconclusive? When dealing with minuscule amounts of ancient fossil bone, one finds that the results of tests are often not as clear as they are with modern material. This fact becomes more and more problematic the farther back in time one goes.

When Mary set out to look for collagen, she attacked the problem on several levels with several different methods—the scanning electron microscope, the transmission electron microscope, atomic force microscopy, mass spectroscopy, and immunoassays. She and her collaborator, John Asara at Harvard, used all of these techniques to pin down what had first appeared to be collagen.

One of the reasons so many techniques were necessary was the minimal amount of organic material in the fossil bone. The traces of proteins were not at all easy to detect, and the threat of contamination or misreading was always present, so every avenue of investigation had to be pursued and every bit of evidence collected.

With the scanning electron microscope, coupled with X-ray scattering, she looked for indications that the substance was collagen. Then she did the same with the transmission electron microscope, which, however, requires samples that are only seventy nanometers thick. The transmission microscope, Mary said, "required that your sections be thin enough for electrons to pass through completely so the electrons are transmitted from one side to the other, which gives you an incredibly high resolution."

"One thing we know about collagen, the major protein in bone, is that it's cross-banded. It has to do with the molecular makeup of the collagen. You get a sixty-seven-nanometer re-peat banding.

"So it's diagnostic of collagen. If you've got a fibrous mate-rial with sixty-seven nanometer bands, you've got collagen."

With the transmission electron microscope scientists can analyze the molecules much more closely by doing what is called elemental analysis. With a scanning electron micro-scope, Schweitzer says, you can tell that dinosaur bone has cal-cium and phosphate with some other traces. "What that can tell me is, yeah, it's a carbonate mineral."

"With the elemental analysis that I can do on a transmis-sion electron microscope. I can tell you that it is one hundred percent fluorapatite. I can tell that it's hydroxyapatite. I can tell you that it is biogenic hydroxyapatite."

Schweitzer also uses atomic force microscopy, in which the researcher can literally push the probe of the microscope phys-ically onto the material and get a quantitative measure of its elasticity. "We can see what the springiness of the material is, compare it with modern bone, and get at how much of the original functionality of the material is there."

The process of gas chromatography requires its own ex-perts to run the experiments. The equipment that Mary uses at North Carolina State—a gas chromatograph coupled to a mass spectrometer—has its own room and takes half a day just to calibrate. The whole process of testing a substance demands the marriage of the most advanced computer technology and expertise with skills that would have been familiar to an alche-mist in medieval Europe.

The preparation of the sample is the kind of work that might have been done by Merlin. It begins with a small amount of powdered fossil that has been prepared by Mary's lab. This is measured into a tin boat, a small open container of tin about half the size of a Chiclet that is shaped something like a gravy boat. The boat is pinched shut and this spitball-sized object, weight precisely determined, is dropped into the well of the chromatograph, at which point the temperature flashes to seventeen hundred degrees Fahrenheit, atomizes the substance at hand, and turns the elements into gases. Those gases are drawn though filters, and molecules of different size travel through the filters at different speeds. Sensors record the amount of the different elements present, and the finding can distinguish the relative abundance of carbon and nitrogen, so you have a carbon-nitrogen ratio, which tells you whether there may be a protein in the sample. It can't prove that the ratio is the result of a protein, but it can detect ratios that exclude the presence of a protein.

Other tests are necessary, such as the immunoassay. Immunology is one of the most sophisticated modern tools for finding out what proteins are in any given substance.

It is based on the amazing ability of immune systems to respond to any sort of foreign invader—or antigen—and quickly create a designer defending molecule—an antibody. Immune system cells are able to shuffle genes like a deck of cards to keep coming up with new hands, except the hands are molecules of different shapes. Every foreign molecule has areas on it, parts of its shape, on which a protein can latch, in a lock-

and-key fashion. When an invader enters the body, whether it's dust or a virus or a bacteria, the immune-cell factories start churning out antibodies of various sorts.

"You can actually take your body and inject it with material from Mars that no human being has ever seen and your body will make anti-Mars antibodies that are specific because of the flexibility of the immune system.

"So we take our dinosaur tissues that we have demineralized. We embed them. We section them very thin, and then we hit them with an antiavian collagen antibody. . . . If the antibody sees something it recognizes, it binds to it. We use a second antibody that recognizes the first one that has bound to tissues, and that second one has been tagged with a fluorescent label. We then put the section under a fluorescent microscope, and if it lights up it's there and if it doesn't light up, there's nothing there."

In the April 2007 issue of *Science*, Mary and six other scientists, including myself and John M. Asara of Harvard Medical, reported finding collagen in B. rex. She offered multiple lines of evidence, including electron microscopy, antibody tests, and mass spectrometry.

In the same issue Asara, along with Mary and three other researchers, reported actual protein sequences, the sort of information than can provide valuable evolutionary connections, in mastodon and *T. rex* fossils. The sequences showed the predicted similarity to bird collagen.

Mary also used material in old dinosaur and mammoth fossils to create antibodies, a reverse process. You inject material from a dinosaur bone into a rabbit, to see if it is in good enough shape to prompt creation of an antibody. Then you see what components these antibodies find and latch on to in modern bone. If, for instance, the dinosaur antibodies recognize collagen in a chicken, that is pretty suggestive that the dinosaur bone of sixty-five million years ago retains collagen that is an awful lot like chicken collagen.

That finding hasn't gone unchallenged. Mike Buckley from the University of York, in England, and about two dozen other scientists, including Peggy Ostrom, criticized the findings in January 2008. In particular they argued that the evidence for *T. rex* wasn't sufficient to conclude that surviving collagen had been sequenced. The evidence was, however, convincing for the mastodon tests, they said. One of the reasons was that, in their reading of the results, it appeared that more change had occurred to the mastodon fossil over a relatively short time than had occurred in the *T. rex* fossil over sixty-eight million years. And they thought the fragmented sequences claimed for *T. rex* showed a similarity to amphibian proteins that didn't make sense.

Asara and Mary replied in the same issue including the technical comment. They cited the many supporting lines of evidence that collagen had survived, and the extreme unlikelihood of amphibian contamination, since amphibians are not native to the Hell Creek area and weren't present in the labs where the substances were tested.

In answering the challenges to their work, Asara and Mary noted that the samples were not tested at other labs because few exist. They do plan to offer material to other labs in the future if they have enough material. But they defend the validity of their tests and the evidence for collagen survival.

Such challenges are essential in science. Mary is convinced, and I am too, that she has the goods on collagen from B. rex, but I would be disappointed and worried if there were no strong critiques.

In April 2008 Chris Organ at Harvard and several other authors, including Mary and Asara, followed up the initial papers by analyzing the sequence data of *T. rex* and mastodon collagen. Chris is a former student of mine. As an outgoing and outstanding graduate student in Bozeman, he had done his dissertation on the biomechanics of dinosaur tails—how they affected movement. He did the research on collagen as a postdoctoral student at Harvard, where he has moved more into molecular biology. That research on collagen found that, as expected, *T. rex* is closely related to the chicken, and a distant cousin of an alligator.

These conclusions are much muddier than they sound, because the evidence is seldom simple and straightforward. It would be wonderful if we could simply pull a chunk of collagen out of a fossil bone and say, "There we have it." Instead we run many tests on a fossil bone, tests that can rule out the existence of a protein, or tests that show direct evidence of collagen. And we interpret the results of the tests, interpretations that are always up for revision and discussion. What we can say, after all our tests, is this: The best explanation of our results, as of now, is that bits of protein have survived for tens of millions of years. It's kind of an opening salvo in a scientific discussion. And, if we're lucky, it will result in more experiments by other people and either the confirmation of our finding, or the development of solid contradictory evidence that tells us we were wrong.

A good question is one that will push our understanding forward when we try to answer it. Can protein molecules survive in original form, in good enough condition to be sequenced over sixty-eight million years? Let's find out. A good question is not always the most profound. It is one that we have the ability to answer. Why is there something rather than nothing? I couldn't tell you and I don't know how to go about pursuing an answer. Whether protein molecules can survive sixty-eight million years is a good question, and we have our provisional answer, which brings up many more questions. How much of the protein can we read? How can we find more fossils like this? How does a molecule survive so long? Why proteins and not DNA? Can we find other proteins? Can we find them from many extinct animals?

"I think the more we study this bone matrix—and eventually the blood vessels and cells, which is where I want to go next—the more information we're going to get on the process of fossilization, the process of degradation, the process of molecular aging, which has a lot of side implications that I think are very intriguing."

Mary's findings have changed how we do field paleontology and I think will have a bigger effect in the future on how everyone does paleontology. We used to collect only the bones and were conscious of the shape and structure of those bones, the gross morphology of animals from the past. In order to prevent disintegration we would immediately coat everything with a preservative.

This was a bit like varnishing the bones, the way you might varnish the wood on a boat to protect it from the elements. It works well to keep old fossils from further disintegrating. But, like varnish, the preservative seeps into the dry fossils, which absorb the chemicals. There is no point in looking for traces of biochemicals from tens of millions of years ago in a fossil bone that has been absorbing new chemicals.

Each summer field season teams from the Museum of the Rockies will now be looking for fossils that have been buried in rock deep enough to make preservation of biomolecules more likely. And we will be making sure Mary gets them right away.

Last year when we excavated the leg of a *Brachylophosaurus* (a duckbill dinosaur), and sent the samples to Mary, it was discovered that there had been some degradation of the sample, even in the short time it took to get the sample to North Carolina. So, to reduce the degradation time, we have taken the lab to the site.

Mary's mobile lab is the trailer part of a tractor-trailer, or eighteen-wheeler. But instead of being filled with freight, the trailer is outfitted as a laboratory. The geology department at North Carolina State University purchased the lab, which was built by the army to be used on a Superfund site, and paid for its transportation to Montana. It has a diesel generator, fuel tanks, water tanks, office space, and a bathroom. It originally cost $500,000 for the army to create the lab. We had it pulled to

Bozeman and the museum, then put about $25,000 more into renovations to create a clean-lab where we could extract soft tissues. The lab has a fume hood, a couple of microscopes, a pure-water system, and other analytical equipment. We hope to get a scanning electron microscope in there eventually.

The lab provided some terrific results this past summer. Mary has some nice material from the Judith River Formation, dating to the Upper Cretaceous, a few million years older than the material in the Hell Creek Formation.

Mary is particularly excited about some of the new material. "We got a lot of great specimens," she says. "We're just learning so much about how we treat the bone in the field. For the first time we started looking at teeth, and those are pretty exciting."

What causes fossil degradation over time and what makes it happen faster or slower are important questions. Mary points out that learning how molecules age could give us new insight into the process of aging in living animals. And there is nothing of more interest to most of us than our own aging. The nature of molecular aging also has implications for our search for evidence of life on other planets. If we know how molecules fall apart, at what rate, under what conditions, we will have a better idea of what we're looking for on Mars or Titan, or beyond.

There is a potential treasure trove of information on evolution. Currently, rates of evolutionary change and the points in the history of life where a lineage diverges have been estimated by the structure of bones—the gross morphology—and by comparing the genes of living creatures to see the degree of difference from one species or genus to another. With that information the pace of evolution can be estimated and evolutionary events backdated. With evidence of protein sequences from ancient creatures, we may be able to dip directly into deep time to test our ideas about evolution.

But there is only so deep you can dig into a dinosaur bone, only so far you can go with the bits of collagen and other biomolecules that may be left. We may not be there yet, but at a certain point we are left with dust in our hands, wondering where to dig next. The answer is: the genes of living animals, because a record of evolution is to be found there. For our purposes the most important record is in the genes of the only remaining dinosaurs—the birds.

4

DINOSAURS AMONG US

CHICKENS AND OTHER COUSINS OF *T. REX*

According to our leading scientists, I am not yet extinct, and they ought to know. Well, there's no use crying about it.

—Will Cuppy, *How to Become Extinct*

Every morning, the dinosaurs make such a racket. I can hear them outside my bedroom window, singing the dawn chorus. When I leave the house they are everywhere. I see them in parks, patrolling the parking lots of shopping malls, on the prairie, along rivers, at the sea, and in New York City, where they live in astonishing numbers. I often find them on my plate at fine and fast-food restaurants.

I'm talking about avian dinosaurs, of course, warblers, starlings, catbirds, cowbirds, robins, orioles, gulls, vultures, kingfishers, sandpipers, falcons, pigeons, and chickens, billions of chickens. I've been saying for most of the book that the dinosaurs never did go extinct, that birds are dinosaurs, descended from theropod dinosaurs, related to *T. rex*, and with a great library of dinosaur genes in their genome.

This is the consensus of scientists now, but it has not always been so, and since the connection of birds to dinosaurs—both in what we have found so far and in what we hope to find—is at the center of the story I want to tell, it is worth stepping back from the digging and pause, before we dive into laboratory work, to do a little evolutionary bird-watching. Our understanding of the relationship of the blue jay to *Velociraptor*, of the chicken to *T. rex*, has itself evolved. It's a good story within a story, the evolution of birds and how we have uncovered it.

We have always known that there was a connection between dinosaurs and birds. Dinosaurs are reptiles and birds clearly descended from reptiles, but exactly which reptiles, and how and when that descent occurred, has been an intriguing puzzle. Modern birds are as magical as any creatures on earth. They are beautiful and clever and they live right in our midst. Unlike almost all other wildlife that we might want to observe, birds do not hide from us. Robins hop across our lawns, gulls chase our boats and congregate at beaches, dumps, and the parking lots of fast food restaurants. Red-tailed hawks sit, unconcerned about the traffic, by roadsides. Hunted birds grow wary, but so many others are so much with us that they have become like the trees and flowers and sunlight. And they fly. That is the single most impressive and intoxicating fact about birds. They fly.

They straddle the winds and stroll the updrafts as if air were solid ground or ocean swells. Intuitively, that puts such a vast distance between them and nonavian dinosaurs that it seems odd to connect them to ancient animals we imagine often as thundering through Cretaceous swamps and coursing across the ancient plains.

How did it occur to us that they might be dinosaurs? How did we know they are reptiles? How did we find out about their evolutionary heritage? In short, where do birds come from?

The answers won't be found in the Hell Creek deposits. By the time B. rex was prowling the Cretaceous delta in the shadow of the Rocky Mountains, the sky, land, and sea were well colonized by birds. Some would seem strange to us now. Diving birds up to four or five feet long with teeth, tiny forelimbs, and short tails fished in the inland sea. They had, for company, the recognizable ancestors of modern birds, including shorebirds, parrots, and flying and diving birds like petrels. Amid these birds and the dinosaurs were the Alvarezsaurids, initially thought to be very primitive birds, now thought by many to be birdlike dinosaurs. No doubt this difficulty we have in pinning down what category we want to put the Alvarezsaurids in did not bother them as they ran about, catching small mammals or other prey. Another bird present in the late Cretaceous was *Ichthyornis*, about a foot long, a diver, with teeth, but with wings long enough to fly.

The great radiation of birds into the many and varied creatures we know today took another ten million years to begin, after the nonavian dinosaurs disappeared. But the birds were already ancient by the time the tyrannosaurs appeared. For their origins we need to delve much deeper.

The proposed ancestors of birds have been many, including turtles, pterosaurs, and other ancient reptiles. In the later nineteenth century, according to Luis Chiappe in *Glorified Dinosaurs: The Origin and Early Evolution of Birds*, several scientists, starting with Karl Gegenbaur in Germany and including Thomas Huxley in England and Edwin Drinker Cope in America, argued for a bird descent from dinosaurs.

Then other reptiles became more popular candidates for bird ancestry. Birds, after all, seemed so different from dinosaurs. Dinosaurs were cold-blooded, sluggish, small-brained, plodding reptiles. Birds are vibrant, quick, and generally have their wits about them. They are engines of heat. Birds live in some of the coldest environments on earth, precisely because their internal temperature regulation is so sophisticated. Owls and falcons populate the Arctic. Skuas and penguins thrive in the Antarctic. Small terns migrate from pole to pole each year in one of the planet's great marathons. Bernd Heinrich, who has studied ravens in the Maine woods, has written eloquently of the gold-crowned kinglet, which lives on the very edge of disaster in terms of energy management. In the North American conifer forests the tiny bird survives the fierce winters by eating constantly during the day, just to gain enough calories to stay alive through the night. At that it has to drop into a torpor of some sort to conserve energy. The gold-crowned kinglet just does not fit the idea of dinosaurs as sluggish and cold-blooded, which predominated for decades until the 1970s.

But our view of dinosaurs changed as our knowledge of birds increased. One of the scientists who helped change our view of both dinosaurs and birds was the late John Ostrom, one of the great paleontologists of the twentieth century. Two discoveries were of key importance.

A NEW VIEW OF DINOSAURS

In 1964, Ostrom, of the Peabody Museum of Yale University, found some unusual bones at a site in the Cloverly Formation near Bridger, Montana. The site, which came to be called the Shrine, was south of Billings, about halfway to the Wyoming border. The site, dating to about one hundred twenty million years ago, in the early Cretaceous, had been excavated once before by Barnum Brown of the Museum of Natural History around 1930.

For the next few years, through the field season of 1967, he and his crew collected more than a thousand fossil bones representing at least three individuals of a new dinosaur. They stopped work because the fossil finds were decreasing and the rock that had to be removed was getting harder and deeper. The dinosaur, which Ostrom named *Deinonychus antirrhopus* (literally "counterbalancing terrible claw"), was named for its two most striking features, a long, stiff tail, and recurved, slashing claws on each of its hind feet.

In 1969 he published a paper naming *Deinonychus* and describing the kind of dinosaur it was—fast, smart, with slashing claws. Ostrom portrayed it as a quick, fierce animal that was smart enough to hunt in packs and had a metabolism that could support sustained effort. It was likely to have been warm-blooded, Ostrom argued, meaning that, like birds and mammals, it could regulate its body temperature separately from the temperature of its environment. Reptiles like turtles, lizards, and alligators depend on the outside temperature to warm them up and cannot function when the temperature drops. At least this is the simple version of what was the common view of science at the time, which was that reptiles did not have independent regulation of bodily heat to any significant degree. Dinosaurs were undoubtedly reptiles, but they did not fit this picture.

With *Deinonychus* Ostrom helped start a revolution in our understanding of dinosaurs, a revolution that I became swept up in, and was able to contribute to, with finds like colonial nesting grounds that also suggested that dinosaurs were unlike the animals we had imagined up to that point.

Ostrom came at the dinosaur/bird connection from both ends. Shortly after *Deinonychus* he made another remarkable discovery, this time in a museum. He found a misclassified specimen of *Archaeopteryx lithographica*, the most famous ancient bird, and the one that produced the most famous fossils, remains in fine-grained limestone that have the quality of masterful etchings. The fossil has the name *lithographica* precisely because of the German limestone deposits, a source of superb material for lithography.

The first fossil skeleton of *Archaeopteryx* was discovered in 1861. It shows us, as Chiappe describes it, "a toothed, crow-sized bird with powerful hand claws and a long bony tail." It was the oldest, most primitive bird known when the fossils were first found, and it still is. That first specimen was sold to the British Museum of Natural History and it is still there. A nearly complete skeletal impression of a comparable fossil of

Archaeopteryx was found in 1877 in another quarry not far from the town where the London fossil was found.

Archaeopteryx is dinosaurlike in many ways. But of course it had abundant feathers, which marked it as a bird immediately. Had it been known at the time that other fossils that were clearly dinosaurs had feathers, the classification might not have been so obvious, since it has many characteristics that make it far different from modern birds, not the least of which are its long tail and teeth. Its skull is reptilian. It is a mixture: long tail, but not as long as its ancestors', primitive spine but not as primitive as those of earlier dinosaurs, and claws at the end of its wings. But it was clearly a bird or a transitional animal between birds and reptiles. Today it is considered a bird, and the earliest bird fossil we have, but not the first bird ever. The study and interpretation of bird fossils show that there must have been earlier birds.

Ostrom made his find because he was working on the origin of flight, and it was for that reason he wanted to examine a specimen of a pterodactyl in the Teyler Museum in Holland. As Pat Shipman describes in her book *Taking Wing,* once the slab in which the fossil was embedded was brought out to him "he carried the slab over to the window where the light was better. In the next instant the oblique sunlight illuminated the slab and brought up the impression of feathers." He knew right away what it was.

The *Archaeopteryx* fossil, misclassified until then as a pterodactyl, was a powerful reminder of how close dinosaurs and birds were. Shipman writes, "As a consequence of this two-part discovery, Ostrom began to revive Huxley's dinosaur hypothesis of bird origins. Birds, he argued with the passion of a sudden convert, are so like small theropod dinosaurs that an unfeathered early bird specimen could easily be mistaken for such a dinosaur."

I first met John Ostrom in 1978 when he came to Princeton, where I was working as a preparator, to talk to my boss, Don Baird, about footprints in the Connecticut Valley. Bob Makela and I had already found the first fossils of baby duck-billed dinosaurs, a complete surprise to the world of dinosaur science because baby dinosaurs were almost never found, and their rarity was a disturbing puzzle. Ostrom looked into the lab where I worked, and commented on the tiny sizes of the baby duckbills. We talked about duckbills and their skulls, in particular about whether bones in their skulls moved when the animals were feeding. Both bird and lizard skulls have this feature, cranial kinesis. John had written that duckbill skulls were akinetic, like those of crocodiles and alligators, and I had presented some evidence at a conference that the duckbill skulls were movable, like those of birds and lizards.

Over time we became friends and in 1995, I invited John to join the Museum of the Rockies crew in the field in Montana to see what we had been doing with his *Deinonychus* site. In 1993 I was interested in the life histories of dinosaurs, particularly whether they had lived in social groups. All of the other sites we had explored were of herbivorous dinosaurs, the prey. With John's permission I sent a field crew to reopen the Shrine. The operation required removal of hundreds of tons of hard rock, using jackhammers, picks, and crowbars, and in the end, very few bones were found, confirming John's good judgment

in not continuing to attack the site. The crew did find important fossils that have led to a much better understanding of
what the skull of *Deinonychus* actually looked like, but at quite
a price. And the quarry gave me precious little information
about the social behavior of *Deinonychus*.

John, like many of the earlier generation of paleontologists,
was a gentleman with a wonderful social presence. Although
we had become friends—and he had been extremely compli-
mentary about my work and an earlier book about finding the
skeletons of baby duckbills, *Digging Dinosaurs*—I still considered him a great scientist, an inspiration, paleontological royalty. So it felt like a privilege to guide him around the site of
the discovery for which he was probably best known. It was
bittersweet, because he was aging, and one of the deepest sat-
isfactions for any dinosaur scientist was slipping away from
him, the prospecting, the excavation, the time travel by shovel
and pickax and jackhammer. It was on his mind as well. After
we toured the site, and John had seen our excavation, he told
me he didn't think he would be venturing out into the field
anymore. Then he gave me his hat.

I can't say hats are as precious to paleontologists as they are
to Texans, but they can be something of a signature, or talis-
man. Think Indiana Jones, without the bullets and Nazis and
special effects. Excavations are never, ever done in the shade.
Where there is erosion and exposure, there is inevitably sun,
and a hat, which is absolutely necessary, can gather memories
and significance. John Ostrom's hat is on the wall of my office,
where it will stay. He died a few years later and we heard the

news when we were in the field. The whole crew was shaken.

THE DESCENT OF BIRDS

John did not just discover an unusual dinosaur, he made a comprehensive argument supporting the descent of birds from dinosaurs. In a 1975 article he summed up other views and presented his own argument with evidence to back it up.

First, he noted that the idea that birds were descended from reptiles had long held sway. "Over the years, several different reptilian groups have been suggested," he wrote, "but for the past fifty years or more the general consensus has placed the source of birds among a group of primitive archosaurian reptiles of Triassic age—the Thecodontia."

The thecodonts were the precursors of crocodilians, pterosaurs, and dinosaurs. They were land animals that had succeeded some of the huge amphibians that evolved as animal life exploded in its colonization of the land. But, John argued, presenting a thorough and detailed analysis of the fossils of *Archaeopteryx*, that were available, this bird fossil was so similar to theropod dinosaurs, specifically the gracile, swift, and predatory coelurosaurs, like *Deinonychus,* that the line of descent to birds was obvious.

He noted similarities in the vertebrae, the forelimb, pelvis, hind limb, and a bone called the pectoral arch. He also dismissed the idea that lack of clavicles or collarbones in theropod dinosaurs meant they could not have given rise to birds, in which right and left clavicles have fused to become what we call the wishbone. Ostrom pointed out that clavicles had indeed been found in several dinosaurs, and that even if they had not been found, negative evidence is never conclusive. Given the rarity of fossils, absence of a characteristic only proves that we haven't found a fossil with it, or we haven't noticed it.

In fact, he concluded, the only characteristics that made *Archaeopteryx* a bird were its feathers and its wishbone. He did not believe *Archaeopteryx* could fly, and suggested that feathers had evolved for insulation, anticipating that other, nonavian dinosaurs would have evolved feathers. Without those two characteristics the skeleton would have been classified as a theropod.

Now is a good time to tackle how such classifications are made. When Ostrom was publishing his work he was tracking descent, a fairly straightforward idea, which led to evolutionary trees much like family trees. Instead of parents and great-grandparents, you would have parent species or genera and great-grandparent species or genera. But genealogy and phylogeny were both alike in that they traced actual descent, trying to establish who fathered cousin Fred and what particular genus of dinosaur gave rise to the first birds. They were, in effect, using the same charts.

Gradually this has been supplanted by cladistics, which is significantly different—even revolutionary—in how it changes the way we think about the past. Cladistics is used not to track ancestors, as in genealogy, but as a way to look at the changing characteristics of organisms over vast stretches of time. It abandons the search for a specific ancestor to any species or genus. Instead it tracks evolutionary change by looking for new characteristics, like feathers or hair or walking on two feet.

A cladistics diagram, or cladogram, starts out with very large groups that share very basic characteristics. Branches appear when new characteristics evolve. These are called derived characteristics because they are derived from a more basic or primitive state. Vertebrates are a very large clade including all animals with backbones. Within that clade are mammals, which have backbones, but also have derived characteristics that they share only with other mammals, hair and mammary glands. Evolution can be tracked from the largest to the smallest clades, as life explodes in diversity and new characteristics keep popping up.

The differences between this approach and older approaches are subtle and profound. Instead of looking for the specific ancestor of birds, for instance, what we try to do is to look at the characteristics birds share with other groups, like the dinosaurs, and what new characteristics they have. There is quite a bit of judgment involved in making sensible groups, or clades, based on specific characteristics. But the close study of old and new traits makes the classification of birds as dinosaurs unavoidable. For example, some of the characteristics that we might think of as being exclusive to birds, like the wishbone, feathers, hollow bones, and oblong eggs, are found in dinosaurs, where they evolved first. There are many more shared features, but most are obscure, like the shape of the wristbone that allows a bird to fold its wings to its sides. If we were to try to do a similar motion with one of our arms, we would have to be able to bend our wrists to the side, rather than front-to-back.

The way we track evolution, shared characteristics like feathers or particular shapes of wristbones mean a common ancestor. Of course, this commonality may be so broad that it is not very helpful. All organisms that have cells with nuclei share a common ancestor, but a characteristic that is shared by ants, falcons, and corals doesn't give us much information about evolution. When groups share a great number of characteristics, then that means they have a common ancestor not very far back. Birds share almost every characteristic that we have noticed with a group called the dromaeosaurid dinosaurs. In fact, it's hard to tell them apart. And birds share so many more characteristics with dinosaurs than they do with other groups that are candidates for avian ancestors, like archosaurs, that we put them in the dinosaur clade.

Cladistics is merely a tool, however, a way of thinking about and categorizing fossils. It is the fossils themselves that are the source of information and, sometimes, exhilaration. In the past fifteen years a series of finds by native and foreign paleontologists in China have produced shock waves of excitement about the ancestry of birds and the nature of dinosaurs.

FEATHERED DINOSAURS

In the mid-1990s one of the best-preserved dinosaur skeletons ever was found in China, in early Cretaceous sediments that provided an extraordinary record of all sorts of life. Three scientists, Pei-Ji Chen, Shi-ming Dong, and Shuo-nan Zhen reported in *Nature* in 1997 finding two skeletons of a chicken-sized dinosaur with the longest tail of any theropod dinosaur and a very large and strong first digit, perhaps a killing claw. The preservation was so striking that internal organs and a last meal of a lizard, as well as two eggs about to be laid were found in one specimen. Most remarkable, however, was the preservation of skin and filaments that the scientists identified as feathers. The dinosaur was named *Sinosauropteryx prima*. (At first it was thought to be a bird.) It was very similar to *Compsognathus*, a dinosaur that early on was thought to be an ancestor of birds. Sinosauropteryx was a coelurosaur, a kind of dinosaur close to birds, in fact the group that includes birds in current thinking.

Shortly thereafter, two theropod dinosaurs with clearly defined feathers were found in the same geological formation in northeastern China that yielded *Sinosauropteryx*, in Liaoning Province. These two dinosaurs were found by two Chinese paleontologists, Ji Qiang and Ji Shu-An; one Canadian, Philip J. Currie of the Royal Tyrrell Museum in Alberta; and one American, Mark Norell of the American Museum of Natural History in New York. These dinosaurs, named *Protarchaeopteryx* and *Caudipteryx*, had both downy feathers and longer branching feathers similar to those in modern birds.

These finds removed feathers as one of the defining characteristics of birds. The two dinosaurs were both classified as maniraptorans, the kind of dinosaur thought to have given rise to birds. More discoveries followed, including feathered dromaeosaurids, another kind of theropod dinosaur. One of the most surprising of these was one called *Microraptor,* a dinosaur about three feet long, with feathers on all four limbs and hind feet that seemed adapted to perching. It certainly looks like it was a tree-living glider and offers considerable support for the idea that flight evolved from dinosaurs that lived in the trees. Xu Xing reported the find in 2003.

Richard Prum, an ornithologist and evolutionary biologist at Yale, who has studied the evolution of feathers, wrote in *Nature* in 2003, in the same issue as the report of *Microraptor,* that the origin of dinosaurs was a settled question. "Birds are a lineage of dinosaurs, and are most closely related to dromaeosaurs and troodontids." With *Microraptor* apparently being a gliding dinosaur, Prum wrote, "there remain no major traits that are unique to birds—with the possible exception of powered flight."

This brings us back to our original and overriding purpose, to build a dinosaur. Since, as Prum writes, powered flight is the only trait unique to birds, we can see quite clearly that causing a bird to grow up as a nonavian dinosaur crosses a thin boundary that grows less clear the more we know. Only small skeletal traits would distinguish a nonavian theropod dinosaur with feathers from an avian dinosaur with feathers. In real terms, however, what I want to see is quite clear—a feathered, running theropod with a tail, teeth, and forelimbs with usable claws. I could put it another way. I want to have a chicken grow up so that we can't tell whether it is an avian or nonavian dinosaur. That would certainly constitute rewinding the tape of evolution.

One objection to this version of the evolution of dinosaurs is that the fossils of theropod dinosaurs with feathers are not older than *Archaeopteryx,* the first known bird, so it doesn't make sense to pick them as ancestors. Clearly those particular dinosaurs were not ancestors of a creature the same age as themselves, but that is not the point. Some of the feathered theropods show primitive characteristics that indicate that their group evolved before birds did.

The much loved duck-billed platypus might help make this clearer. The platypus is a favorite of children, evolutionary biologists, and the sort of person who likes to throw the word *monotreme* into the conversation. A monotreme is a very ancient kind of mammal that lays eggs. There are only two of them: the platypus and the echidna (spiny anteater), both of them native to Australia. They are oddities among the odd, since Australia is set apart from the rest of the world by having no native placental mammals except human beings. Kangaroos, koalas, and the rest are all marsupials, with protective pouches for their tiny young to continue their development until they are ready to face the outer world. Placental mammals like us give birth to fairly well developed young that survive outside the mother's body.

But the platypus is something else again. As Ogden Nash, who might be said to have his own evolutionary branch among poets, wrote, "I like the duck-billed platypus, Because it is anomalous." It has a duck's bill, more or less, and lays eggs, but it has mammary glands, although no nipples. The young, once they hatch, must suck the mother's milk through thin skin over the glands. To top it all off, the platypus has venom, delivered by spurs on its legs. It seems like a cross between a mammal and a reptile and, unsurprisingly, its genome has what we think of as reptilian and mammalian characteristics.

Most of us like the platypus for the same reasons as Ogden Nash, but it has evolutionary importance because we think the first mammals probably had some of these reptilian characteristics such as egg laying. So the platypus has primitive characteristics that were lost in other mammals as time and evolution proceeded. But evolution is not restricted to one line. At the same time that mammals with what are called derived characteristics, such as nipples, were evolving, and other animals like the platypus were disappearing, the platypus survived. In the same way, single-celled life-forms did not disappear or stop evolving as multicellular animals appeared and diversified.

So when we look at fossils we try to identify primitive characteristics, and derived, or novel, characteristics. Naturally, the novel characteristics appear later in time. And our knowledge is always changing and developing. At one time we thought that feathers were a derived characteristic that identified birds. No longer, since we know of nonavian dinosaurs that had feathers. A number of such fossils have been found in China.

The evolutionary path to birds is now seen as follows. The first dinosaurs emerged in the Triassic, about 225 million years ago, from reptiles called thecodonts, and split into two sorts, ornithischians and saurischians. Here the terminology is a bit misleading, because although the ornithischians are named for birdlike hips and the saurischians for lizardlike hips, the birds arose within the saurischian lineage. The saurischians split into sauropods, like the big, long-necked herbivorous brontosaurs, and the theropods, carnivorous dinosaurs. Birds are theropods, and although we don't know which theropod gave rise to them, it was small, fast, smart, and carnivorous. The best guess is that birds arose from primitive coelurosaurs, which are first known from the early Jurassic, between 175 and 200 million years ago.

Archaeopteryx is, however, the first known bird. It emerged around 150 million years ago. After it we can trace bird evolution through several steps. Modern birds appeared about 55 million years ago, and within those, the galliform birds appeared about 45 million years ago. The domestication of *Gallus gallus,* the red jungle fowl that became our domestic chicken, apparently began around 5,000 years ago.

Remarkably, it is in this genome, 50 million years removed from its nonavian theropod ancestors, that the information resides to grow a dinosaur. I mentioned earlier one of the most recent calls for changes in paleontology, by three scientists, including Kevin Peterson at Dartmouth. It summed up progress to that point in merging paleontology and molecular biology and pointed the way for much more mixing of the two disciplines, in the new, hybrid field of molecular paleontology. He and his colleagues pointed out that there is a vast repository of molecular fossils within the genomes of living animals, and that "we are now in a position, both technically and methodologically, not only to explore this molecular fossil record but also to integrate it with the geological fossil record."

What is particularly interesting to me, perhaps because I agree with them, is their argument that there must be a marriage or merger of the skills and knowledge of molecular biology and paleontology. Neither is sufficient without the other. The skills of the molecular biologist and the understanding of the mechanisms of genetics are necessary, as is an understanding of the fossil record and the grand sweep of evolution and the classification of life-forms.

This is certainly true. And although it may be my bias, it seems to me that it is often paleontology that sets the table and makes possible the questions that molecular biology has the knowledge and skills to answer. That is certainly the case when it comes to dinosaurs and birds. It is in birds that we will find the molecular fossils that lead us to learn more about dinosaurs and their evolution.

5

WHERE BABIES COME FROM

ANCESTORS IN THE EGG

The problem of development is how a single cell, the fertilized egg, gives rise to all animals, including humans. So it really is about life itself.

—Lewis Wolpert, *The Triumph of the Embryo*

A map of the chicken genome, actually the genome of the ancestral chicken, the red jungle fowl *(Gallus gallus),* was published in 2004. The achievement followed on the mapping of a number of other genomes, including, of course, our own. So it did not receive any great fanfare. But this was the first avian genome and it was immediately compared to human and other genomes in a search for insights about the separate paths evolution has taken. The last common ancestor of mammals and birds dates to about 310 million years ago, which is a long time for separate evolution.

And there were a number of intriguing differences.

One major difference is that the chicken genome is one third the size of the human genome, which contains twenty to twenty-five thousand genes. Chickens also have many fewer repeating sections of DNA. Humans, for reasons that are still not understood, have much DNA that has been called junk because it was thought to be leftover and nonfunctional. The thinking now is that much of it is useful, in ways that we hope to figure out as we map the details left out in the original studies of the genome. But birds are more economical in their DNA. Not surprisingly, the chicken also has a specialized set of genes for the keratin that goes into beaks and feathers. There are long sections of DNA that are the same in chickens and humans, but some of these are of unknown function.

Comparison has always been one of the key techniques of science, and using other creatures that are more easily observed has also been a key to understanding human biology. Genome mapping is equally easy in chickens and humans, but studies of genes are only part of the way molecular biologists mine the treasures of modern animals to understand the path of evolution.

Another technique of great importance has been the study of development, of embryology, to penetrate the great mystery of how a fertilized egg—one cell—grows to an adult organism. Today we study how this process is directed by genes, and how it relates to evolution, but development has been studied since antiquity.

Aristotle is considered the father of embryology, if not biology. In 345 BC he observed and recorded the development of the chicken embryo in the egg. As far as we know, he is the first experimental embryologist. In *Great Scientific Experiments* (1981) Rom Harré—a philosopher of science, prolific popular writer, longtime professor at the University of Oxford, and now teacher at Georgetown—examines one of the philosopher's Hippocratic writings, in which Aristotle follows up on an experiment proposed by an unknown author:.

"In the work *On the Nature of the Infant*," Harré writes, "an exploratory study is suggested in the clearest terms. 'Take twenty eggs or more, and set them for brooding under two or more hens. Then on each day of incubation from the second to the last, that of hatching, remove one egg and open it for examination.'"

Aristotle apparently followed this suggestion to the letter. (The unknown author who suggested the experiment never seems to have actually done it.) In the *Historia Animalium* Aristotle recounted the results, providing a source that was relied on for more than a millennium. He described the first hint of an embryo after three days, the development of the yolk, the first hint of the heart, which, he wrote, "appears, like a speck of blood, in the white of the egg. This point beats and moves as though endowed with life."

Even for the modern reader Aristotle's eye for detail and the clarity of his writing are remarkable. "When the egg is now ten days old the chick and all its parts are distinctly visible. The head is still larger than the rest of its body, and the eyes larger than the head, but still devoid of vision." He continued to observe after hatching, and noted that "ten days after hatching, if you cut open the chick, a small remnant of the yolk is still left in connection with the gut."

Aristotle was not pursuing an idle interest, or a particular attachment to chickens. The original proposer of the experiment was writing about human development, and the chicken egg was a means to watch an embryo grow, the presumption being that human infants had to share some aspects of this development. Also, to the Greeks and still to us, the growth of an organism is one of the most profound biological mysteries. It is the child's inquiry writ large as a scientific question that still demands our full attention: Where do babies come from?

The different answers to this question have fallen into two schools of thought that date to the Greeks and still have resonance today. One school favored preformation and the other epigenesis. In preformation the organism already exists in some miniature form in the parent. Everything that is needed for the adult form is already there. In epigenesis, however, the raw material of the new creature is shaped and changed as it develops. So the nature of the individual is largely determined during growth.

To imagine a miniature human being, or chicken, for that matter, already existing in the egg is too simplistic to the modern mind, but the essential philosophical difference still resonates. Is every detail of individual human behavior and personality prescribed in the genetic code? That would be a kind of preformation. You could have a gay gene, or a crime gene. But if genes are more like the notes for a musical composition, but without the tempo or orchestration, or even specifying the instrument that is to play it, then development would be closer to epigenesis. Hormones in the mother's system could affect the development of the fetal brain or the sexual organs, and maternal nutrition or drugs could enhance or harm the development of the embryo. Given the popularity of the idea that playing Mozart to a pregnant mother will be beneficial to the growing embryo, it is clear that the idea of epigenesis still has currency in the modern world.

Science is just now passing out of a period during which genes received so much attention that it seemed all researchers saw biology from a modern preformationist perspective. Now it seems genes don't tell anything like the full story. New research is giving us an understanding of subtle chemical events that affect the expression of genes and the development of the embryo, chemical events not determined by the genes themselves. This is a new kind of epigenesis and so far, no link has been established to Mozart.

Deep philosophical questions about the essential nature of the new individual aside, we have gained a vast amount of operational knowledge over the course of the past two centuries, in particular the second half of the twentieth century, about how the growth of the embryo is orchestrated, about what directs the astonishing unfolding of form we see in a growing fly, mouse, or human embryo. And it is this knowledge, coupled with our knowledge of genetics, that enables us to think that we might be able to change the course of an embryo's development so that it grows more in the fashion of one of its ancestors than in the normal way.

Knowledge of the structure of DNA and the genetic code has helped bring us to this point, but it has also misdirected our thinking in some ways. We know that in the genetic code are sequences that produce proteins and that these proteins are crucial in determining different aspects of a growing organism. We know that given certain genes, eyes will be blue. With others, eyes will be brown. In the fruit fly there are genes for crinkly wings, smooth wings, and no wings. We know of diseases that are caused by a change in a single gene. And we know now of genes that cause or increase the risk of diseases.

Still, much of this is a bit like knowing that if you put two chemicals together there will be an explosion. But what is the mechanism? What determines the force and direction of the explosion? What is the chemistry? Well, that's when things get beautifully complicated.

Over the last quarter century or so scientists have made astonishing progress in embryology, moving toward the goal of being able to write down what would essentially be the "program" for the development of an organism starting with a fertilized egg. That is to say, every gene action and action on a gene that results in growth and development could be cataloged. One would have "the instructions" for a worm, or a fly.

And, in the last twenty years that knowledge of genes and their actions and how they are controlled has been applied to the understanding of evolution. That has given us the field called evolutionary developmental biology, or evo-devo. This understanding can enable us, with a few nudges, to see if we can rewind the tape of evolution from the chicken toward the nonavian dinosaur.

THE EVOLVING EMBRYO

Historically, there has long been an interest in the potential connection between the growth of the embryo—ontogeny—and the evolutionary history of an organism, its phylogeny. This connection was explored in one of the first books by the late Stephen Jay Gould, *Ontogeny and Phylogeny*, published in 1977, when the great importance of regulatory genes in development was just beginning to be recognized. Gould's book circles around a statement familiar to scholars who know the history of the development of evolutionary theory, although it may sound something like scientific double-talk to most people: that is, the claim of Ernst Haeckel, an early student of the importance of embryology in determining the form of organisms, that "ontogeny recapitulates phylogeny." He argued that one can see the evolutionary history of a species repeated in the embryonic development of an individual of the species. More precisely, the embryo passes through the adult stages of its ancestors, showing in compressed time and space the course of evolution that produced it. This was intuitively appealing because anyone can see that a human embryo, for example, goes through stages where it looks like some of our ancient ancestors, like fish and amphibians. There are what appear to be gills and a tail.

The idea is oversimplified, however, and had been long discredited by the time Gould was writing his book. What he did was put the statement in its historical context. He argued more than once that mistakes in science could be as useful and enlightening as correct ideas, sometimes more so. And he treated the notion of recapitulation as a mistake that had more sub-stance and interest than its rejection. However Haeckel went wrong, his idea pointed to an important connection between embryology and evolution that had been pursued by scientists but then had been largely abandoned, to the impoverishment of evolutionary theory.

For one thing, Gould argued, changes in timing of embry-onic development could make dramatic changes in evolution, particularly when different aspects of development followed different schedules. For instance, an evolutionary change would occur if the developmental path to sexual maturity were speeded up but all other sorts of growth stayed at the original pace. If frogs became sexually mature as tadpoles and never made it to the frog stage, producing a new species of adult tadpoles, that would be quite a dramatic evolutionary step.

Something like this occurred with humans, Gould wrote. In our development, which extends long past the embryo, ju-venile stages last much longer than they did in our primate ancestors. Consequently, when we reach sexual maturity we are, physically, at an ancestral juvenile stage. Our mental plas-ticity that enables lifelong learning could also be a juvenile characteristic that stays with us into old age.

Gould saw that changes in regulation of gene expression would be central to any understanding of the mechanisms of evolutionary change. Of "the growing discussion on the evolu-tionary significance of changes in gene regulation," he said, "I predict that this debate will define the major issue in evolu-tionary biology for the 1980s." He continued, "I also believe that an understanding of regulation must lie at the center of any rapprochement between molecular and evolutionary biology; for a synthesis of these two biologies will surely take place, if it occurs at all, on the common field of development."

The synthesis did occur, in the development of evo-devo. Not all scientific disciplines need nicknames, but this one, also called devo-evo by some, was in desperate need of a way to simplify the full descriptor, "evolutionary developmental biology," or "developmental evolutionary biology." Gould did not anticipate rewinding the tape of evolution, however. As mentioned earlier, he wrote that the tape could not be rewound and run again with the same result, although he was not talking about laboratory experiments.

Sean Carroll at the University of Wisconsin–Madison has been one of the pioneers of the evo-devo field and a very effective popularizer. He gives Gould a lot of credit for foresight. And he points out that although the structure of DNA, the nature of genes, and the nature of genetic changes in populations had been well studied through the 1970s, the evolution of the form, the shape, of organisms had not been deciphered. Indeed, he has written, this was largely because the knowledge of embryology itself was lacking. "How could we make progress on questions involving the evolution of form without a scientific understanding of how form is generated in the first place?"

To illustrate how recent the change was, he has written that through the 1970s, "no gene that affected the form and evolution of any animal had been characterized. New insights in evolution would require breakthroughs in embryology."

Those breakthroughs occurred largely among a group of scientists known to themselves and others in related fields as the fly people. That is to say, they defined themselves, and were defined, as is common in science, by the organism that they studied. In this case it was the experimental organism that was the twenty-first century's experimental hero in many studies of genetics—*Drosophila melanogaster*, otherwise known as the fruit fly.

Experimental biology, and in particular the investigation of how genes and inheritance and development all work together, is divided into camps of researchers who work on one animal "model" or another. *Drosophila* is one, the worm *C. elegans* another, the chicken yet another. The mouse is an animal model that has been very useful for testing drugs and for creating strains that are lacking one gene or another, so-called knockout mice. In this manner scientists have been able to cause obesity, cancer, and diabetes in mice, and even to cause effects that are similar to schizophrenia and Alzheimer's disease. All the so-called model organisms are relatively easy to maintain in a laboratory and breed quickly enough for researchers to design experiments that will show the effects of genetic change in weeks or months rather than years or decades.

Organisms become laboratory models because a body of work is built up through laboratory studies and researchers can build on previous work. The result is, with luck, a deep and thorough understanding of one system that can then be applied to others, although up until the end of the twentieth century there was little thought that worms and flies would be as similar to people in their genes and organization as they have turned out to be.

Dr. Thomas H. Morgan, of Columbia University, started *Drosophila* on its career. It was used to study the simple rules of Mendelian inheritance and to help scientists understand the growth of an embryo. And this is where the major discoveries were made in controlling genes that turn other genes on and off and that determine the patterns of growth that govern the development from egg to fly.

This body of knowledge was developed in such detail that it would fill libraries. And it is this work that provided an understanding of development that challenged the standard view of evolution at that time, and which turned out to have a shocking relevance to human biology and even behavior. Genes had become the focus of inheritance and evolution. It was quite clear that DNA contained the information from which a fly or worm or human being was made. The genes were transcribed into RNA, a single strand with a mirror genetic code, that was then run through cellular machinery to produce a protein. As described before in the discussion of Mary Schweitzer's work in looking for ancient molecules, proteins are the molecules that do all the work in the body. An organism is built of proteins, by proteins, for purposes that, so far, are unknown to anyone.

It was also clear that mutations in DNA provided the means for changes in proteins and changes in the external characteristics of organisms. At this time, long before the age of the genome and the comparison of one genome to another, population geneticists and evolutionary biologists worked on the notion that accumulations of small changes (microevolution) led to large changes in species and genus and the outward form of organisms (macroevolution).

Consequently, it was thought that the genes of worms and people, of flies and mice, would have to be very different from one another. Homologous genes, which is to say genes that serve the same purpose in different organisms, were imagined to be rare. Carroll writes, "The greater the disparity in animal form, the less (if anything) the development of two animals would have in common at the level of their genes." One of the architects of the Modern Synthesis, Ernst Mayr, had written that "the search for homologous genes is quite futile except in very close relatives."

This was the mainstream view, and before the research on the development of the fly embryo, it made sense. What the fly research did was first to show how development proceeded, itself a profound and thorny scientific problem, and second, to illuminate the path of macroevolution—how different animal forms had first appeared and how they later evolved into new forms. Such major, visible changes were not brought about by the accumulation of many, many tiny changes. Instead it seemed that evolution was working as a self-assembling kit, with many similar parts. The genes that were most powerful in the direction of evolution were those that determined changes in size, shape, number, and location of the basic parts and when, where, and how the parts were put together.

Although evolution and development cannot be separated, the subjects are so complex that it is necessary to take them one at a time. The magnitude of trying to understand how a single cell grows into a fly, frog, pony, or chicken, let alone a human being, is almost impossible to overstate. Scott Gilbert—in a textbook, of all places—captured the extent of the problem vividly. He was writing about Wilhelm Roux, a founder of the field of experimental embryology, who wrote a manifesto in 1894. Roux's view, Gilbert writes, was that understanding the causes of development was "the greatest problem the human intellect has attempted to solve." That was so, according to Roux, "since every new cause ascertained only gives rise to fresh questions concerning the cause of this cause."

MASTER GENES

It was enough to make a theoretical physicist throw up his hands. And it was the fly that provided the answers. Bithorax was the first of the single-gene mutations in the fly that turned out to be so important in development. The mutation made the hind wings look like the front wings. Others came quickly. "A rather spectacular mutant, antennapedia, causes the development of legs in place of the antennae on the head," Carroll wrote. A number of these genes were discovered. They were called homeotic genes, and in each case a mutation turned one body part into another part, antennae into legs, as with antennapedia, or hind wings into forewings, as with bithorax. Another characteristic of these genes is that all applied to modular body parts, building blocks that when varied in size, shape, or number produced a creature whose form was different. A fly or other insect could have different numbers of wings, or legs, all made from basic, repeatable body parts.

These were obviously very powerful genes governing the overall pattern of the fly body, and, as it turned out, not just the fly body. They were the master genes of evo-devo. There were eight genes in two clusters, the Antennapedia Complex and the Bithorax Complex, five for the front half of the fly and three for the back half. Carroll writes, "Even more provocative, the relative order of the genes in these two clusters corresponded to the relative order of the body parts they affected." In other words the physical arrangement of the genes along the chromosome put them in the same sequence—head to tail—as the parts of the fly they governed. The genes shared a stretch of DNA called the homeobox and they were called homeobox, or HOX, genes. In each of these genes the homeobox contained the code for a stretch of protein that was designed to latch on to other genes in order to turn them on and off. Similar homeobox stretches were found in genes and proteins in frogs, birds, and mammals, meaning that throughout the animal kingdom these HOX genes and proteins were turning genetic switches on and off during the development of the embryo. The HOX genes were master coordinators of development. And they were so similar that it meant they had remained the same over the course of five hundred million years of evolution. That is how far back one would have to go to find the common ancestor of fruit flies and mammals. These master genes were clearly essential to life.

As research continued, in the '80s and '90s other master controlling genes were discovered, genes for building essential organs like limbs and hearts. Genes to control patterns of growth. The way all these genes in what is called the "genetic tool kit" function is by producing proteins that switch other genes on and off (transcription factors). They also produce proteins to travel to other cells and set off sequences of gene activation that alter how the cells behave, how they move, and at what stage of development, and what the rate of growth should be.

The whole structure of control is not fully understood, and some stretches of DNA have been found to produce small snippets of RNA that are never translated into proteins. These micro RNAs also turn genes on and off, genes that may then produce either RNA or proteins to control other genes, and so on down the line. The potential combinations boggle the mind, but if one were able to map out every instruction, every gene activation and chemical event, in order and location, one would then have the instructions for building a worm, or a fly.

Sonic hedgehog, for example, is the name for both a gene and its protein. The protein is a transcription factor that affects growth. You can take a developing embryo and add or inhibit sonic hedgehog without actually changing the genes, and you will turn on or off the growth of a forelimb, or a tail. The odd name is a result of laboratory humor among fruit fly geneticists. One version of the gene causes fruit fly embryos to be covered with spikes so they look like a hedgehog. Sonic hedgehog comes from the cartoon character. Groucho and smurf are other such genes, also death executioner Blc-2.

One of the important families of growth factors observed in developing embryos is the bone morphogenetic protein family, BMP. Different kinds of BMP, identified by number, control genes that cause growth of bone cells. But what is the control for turning on the control gene? As Carroll describes it, "there are separate switches for BMP5 expression in ribs, limbs, fingertips, the outer ear, the inner ear, vertebrae, thyroid cartilage, nasal sinuses, the sternum, and more."

Each switch has different sequences of DNA within it, to which different proteins bind. "An average-size switch is usually several hundred base pairs of DNA long. Within this span there may be anywhere from a half dozen to twenty or more signature sequences for several different proteins." Carroll estimates the different combinations possible with five hundred DNA binding proteins that can work together in pairs or large numbers to activate sequences in switches. There are "12,500,000 different three-way combinations and over 6 billion different four-way combinations."

Perhaps even more intriguing is that, "There is no 'masterbuilder' in the embryo," as Lewis Wolpert wrote in *The Triumph of the Embryo*. The cells talk to each other. "There is no central government but rather, a number of small self-governing regions." And one event determines the next, writes Wolpert, "There are thus no genes for 'arm' or 'leg' as such, but specific genes which become active during their formation. The complexity of development is due to the cascade of effects."

Imagine the development of an embryo as a self-conducting symphony in which the sound of the bassoons triggers the tympani. The bassoons are triggered by the violins, but depending on what the violins play, and when and how loud, the bassoons may play differently, which will affect the tympani. And if the tympani play long enough, that stops the violins.

Embryologists have watched every stage of growth of organisms like the fly and the chicken, and have mapped where cells go to become a brain or a liver, and what chemicals are present in the cells when they proliferate or change. They have looked at limb growth in great detail and noted when the buds that become digits first appear and how many grow and which ones do not grow. They have watched the death of cells functioning to sculpt shapes that then continue to grow.

And it has become clear that this is how the forms of animals change during evolution. A mutation in a master or signaling gene, or a change in a switch, or switches, extends the fingers in a bat's wing and makes the webbing grow. For each new shape and form, there is no new suite of genes that provide a whole new set of detailed instructions for a wing instead of a limb. Changes in regulation reverberate through the system of switches and feedback loops to create new forms.

Charles Darwin's notion of natural selection remains as the most powerful, most fully understood force of nature. It "selects." Some changes in development will be useful, while others will be fatal. But on the evolutionary voyage from dinosaur to falcon, what happens is not that a whole new set of falcon genes is developed for beak, wings, and eyes. Instead the instructions for limbs, feathers, eyes, and tail are changed so that the same building blocks of the vertebrate body are put together in different ways.

The hope for applying knowledge of development to evolution and, for our purposes, to find a way back through the extinction barrier, is to link microevolution to macroevolution. If we can tie development, recorded down to the specific gene and its protein product, to the gross anatomy of fossils, we will have a whole new level of understanding about the evolution of form. This kind of work is in its early stages, but there are some good examples.

HOW FEATHERS GROW

Feathers are one feature, highly pertinent to both avian and nonavian dinosaurs, for which this has been done in elegant and satisfying detail by Richard O. Prum of Yale and several colleagues.

Their work is all the more interesting because, in the absence of evidence from genetics and developmental biology, a theory of feather evolution had been developed that seemed to make sense but turned out to be impossible. "According to this scenario," Prum and Alan H. Brush wrote in *Scientific American* in March 2003, "scales became feathers by first elongating, then growing fringed edges, and finally producing hooked and grooved barbules."

To understand why this couldn't have happened, it's necessary first to understand the structure of that lovely feather floating in the wind, or contributing to the fluffiness of your pillow. Feathers are essentially long tubes with branches. The branches also have branches, and those branches again have something like branches, except that the last twiglike extensions are hooks or barbules that hold the feather together.

There are two different sorts of feathers. One is the blue jay or pigeon feather you may find on the ground, the turkey feather you can buy if you tie flies to catch trout. The other is found in great numbers as the down in your sleeping bag, comforter, or winter coat. The first is pennaceous and the second is—and this has to be one of the great words of biology—plumulaceous. The pennaceous feathers have the branching described above, while the plumulaceous feathers have very little main stem and instead a tangle of lesser branches, with barbules that link together, forming the air-trapping matrix that keeps birds warm, and people as well, in their sleeping bags and puffy mountaineering coats.

The first step in understanding what feathers are and how they evolved was achieved simply by tracking embryonic growth at a microscopic level. Feathers grow out of the skin or epidermis, the outer layer of cells in the developing embryo. Part of the skin starts to thicken, and then to grow out into a tube, while around the growing tube a cylinder of cells form the follicle. The follicle keeps generating a kind of cell that produces keratin, the substance in fingernails and hair. The new cells at the bottom push the old cells at the top, "eventually creating the entire feather in an elaborate choreography that is one of the wonders of nature," Prum writes.

One aspect of the growth is indeed wonderful, and complex. As the hollow tube grows, something happens with the part of the follicle called the collar, which is the source of the growth of keratin-producing cells that push the central, hollow shaft of the feather out from the skin. It begins producing ridges on the central shaft that grow in a helix on the tube, turning into the main branches as the feather grows. Then the barbules grow from these branches. All of this happens at once, which gives a hint of the mystery and wonder in the way organisms grow from one cell to a complex creature, with so many cells forming so many and such complex patterns, all timed to occur at the right moment and directed to the right place. The feather is just one small example of this sort of change in concert.

Prum and other colleagues proposed that in the course of this development they could see the way feathers had evolved. Primitive structures, like those they identified in the early stages of feather development, must have appeared first in evolution. Animals must have existed that had only these tubelike structures. Only later did the feathers that let birds fly emerge.

In other words, feather evolution, like feather growth in the embryo, proceeded by discrete steps. And one step had to be completed before the next one could occur. Each step depended on what had gone before. The final product, the feathers that enable the flight of falcons and swallows, came long after the feather first evolved. And since those first feathers had absolutely no connection to flying, feathers had to have evolved for some other purpose. The feather has been one of the features creationists have long pointed to as an impossibility for evolution. How could feathers, a truly novel development, not just a longer arm or a thicker skull, evolve on their own and just happen to be useful for flying? Prum and colleagues showed exactly how that could and did happen. Features emerged that served one purpose, and as other features were built on them they changed into the structures we see today.

First to evolve were simple tubes, hollow cylinders, then barbs that formed tufts on the tubes. In the next stage feathers became tubes with branches that had tufts, or barbules. There was one more step, which was for the barbules to change shape to have hooks at the end. These hooks are what allow a feather to close and feel as if it is one piece, repelling water, or pushing on air. After the stage of the hooking barbules, the change that produced true flight feathers could have taken place. This was an asymmetric feather, with more on one side of the central tube.

Prum and John F. Fallon and Matthew Harris at the University of Wisconsin–Madison went deeper into development, using techniques to observe which genes were active at which stages and in which locations in the growing feather. They found two well-known genes and the proteins they coded for. Sonic hedgehog and one of the bone morphogenetic proteins, BMP2, were present in different places and different concentrations promoting growth (sonic hedgehog) and the differentiation of new kinds of cells (BMP2). BMP2 was also limiting cell proliferation.

First they would appear where the feather germ was starting, later at the beginning of the ridges that turned into the first branches. The two proteins directed the growth of the feather, and did it in stages, just as Prum and colleagues were suggesting, with each stage possible only because of the one before it. Without the feather germ there could be no ridges or branches or tufts. The general picture they saw in development and proposed in evolution was that first came the central shaft, the hollow tube. Then came downy tufts. Finally came the helical ridges, organized branches, and barbules that made modern feathers, the sort that can be found on a starling or on *Archaeopteryx*.

On a cold night when you crawl into the warm cave under a down comforter, you are taking advantage of millions of years of evolution, mediated by two genes and the proteins they code for—sonic hedgehog and bone morphogenetic protein .

This made the steplike sequence of development clear, but more evidence was needed to link the developmental sequence to an evolutionary sequence. Some of this was readily available in the great variety of feathers in modern birds. Each evolutionary stage of feather development could be seen on some living bird. So, Prum was not inventing any structures that were unknown. All these feather types had appeared on birds at one stage or another.

Nothing they had learned had falsified their hypothesis. Nothing had proved it either. Of course, in historical sciences, like paleontology or evolutionary molecular biology, proof is not possible in the way that it can be obtained in a physics experiment. But predictions can be made and evidence produced that supports or refutes the validity of the predictions. Prum and his colleagues, in describing the sequence of evolution, were, in effect, predicting that extinct organisms existed that had primitive feathers, mere tubes, and downy feathers, and that these should have existed before *Archaeopteryx*.

Paleontology came to the rescue with the discoveries of feathered dinosaurs, which I described in the last chapter, in the 1990s in China. These were just what had been predicted.

As Prum writes, "The first feathered dinosaur found there, in 1997, was a chicken-size coelurosaur *(Sinosauropteryx);* it had small tubular and perhaps branched structure emerging from its skin." Later, other dinosaurs were found with pennaceous feathers. The variety of feathers, including the simple tufted sort that would correspond to the second stage of feather evolution in the Prum plan, all of them on dinosaurs, gave further support to this idea of feather evolution.

Prum's exhilaration in the *Scientific American* article produced one of the great scientific sentences: "These fossils open a new chapter in the history of vertebrate skin." Indeed.

Birds became a subset of theropod dinosaurs. Dinosaurs acquired feathers. *T. rex* may even have had them. The idea of feathers evolving from scales was undermined. Scales don't grow as cylinders, but with a distinct top and bottom. And it became clear that feathers did not evolve for the purposes of flight. Why they evolved we don't know. Nor can we say when they evolved. And we have found that we will probably never be able to say when birds evolved. All evolution in reality is a continuum, with no sharp distinctions. And nowhere is this clearer than in the transition from theropod dinosaur to avian dinosaur. Arguments now exist over whether some of the Chinese dinosaurs are birds.

In the work on feathers Prum demonstrated and articulated the direction that paleontology and evolutionary biology must take: the same direction that others have favored. As he concluded, "Feathers offer a sterling example of how we can best study the origin of an evolutionary novelty: Focus on understanding those features that are truly new and examine how they form during development in modern organisms." In fact, he refers to it as a "new paradigm in evolutionary biology" and one that is likely to be very productive. In a forgivable pun he ends by saying, "Let our minds take wing."

HAND TO WING

Another example of how developmental evidence can be used to infer what happened in evolution has to do with the bird hand. Hans Larsson at the Redpath Museum, McGill University, and Günter Wagner at Yale, along with others, have been occupied with a problem that is obvious on the surface, but leads to murky twists and turns when you start to look at it more closely.

Certainly development of an embryo has some parallels to the evolutionary history of organisms. And it may be tempting, as has happened in the past, to come up with a just-so story of an evolutionary process that would follow the developmental process that we can see. But how does one justify the conclusion? What counts as evidence? What are the rules of logic and experiment that constrain scientists who want to point to the ways the evolution of the feather or bird hand occurred?

For laboratory sciences the problem is simple. It's the old scientific method. You come up with a hypothesis and then use experiments to test the hypothesis. It has to be falsifiable so that it can be proved wrong. This works with microevolution, changes in specific genes that we can see. We could hypothesize that if we put bacteria in an environment laced with amoxicillin, the amoxicillin-resistant ones will live and prosper. The bacterial population will evolve to become untouchable by that antibiotic. In fact, this is an experiment being conducted right now in the ears of American toddlers. A Mississippi of pink liquid amoxicillin flows through the nation's pharmacies and the bacteria that cause ear infections are becoming tougher and more resistant.

We could no doubt find the genes responsible for bacterial resistance and demonstrate evolution in action. But macro-evolution occurs over time. The study of how birds evolved, of where mammals came from, of how primates appeared— these issues have to be studied historically. And here, the logic of science becomes a bit different. There is, of course, no proof in science as there is in mathematics. You can prove something wrong. And you can accumulate evidence in support of a theory until it becomes strong and well-founded. But any theory is always susceptible to new evidence, new theoretical approaches.

Hans turns to two ideas as the philosophical basis for his use of developmental stages in attempting to understand evolutionary events.

One is the idea of forensic evidence. Just as coroners determine the manner of death by looking at a corpse, he writes, so scientists can reason the course of evolution by looking at the fossil record.

Another important notion is at the heart of the reasoning that ties changes in the development of the embryo to changes in the shape of animals in the course of evolution. And that is that for a developmental event, a change in how an embryo grows, to be linked to an evolutionary event, a change in the form of adult animals over the course of evolutionary time, the two events have to be of comparable complexity. There has to be a kind of symmetry.

We can see, in the fossil record, how nonavian dinosaurs gave rise to avian dinosaurs and how those avian dinosaurs, the birds, themselves evolved. Along the way a five-fingered hand changed to a three-fingered hand, changed to three fingers stretched into a wing. And we can see the development of the wing as a chicken embryo grows. If we want to draw conclusions to connect the laboratory and the fossil evidence, we need scientific rules of engagement, a clear understanding of what constitutes scientific proof in linking development and evolution.

The symmetry that Hans has argued must exist between the two events is not the supersymmetry of theoretical physics that holds that for each subatomic particle there is a supersymmetrical "swarticle"—requiring squarks, selectrons, and sprotons, all of which may have something to do with the dark matter that seems to make up most of the universe. No. Evolutionary theory may get complicated, but it is not yet ready to match theoretical physics in its complexity.

The symmetry that Hans is talking about is between cause and effect. In this case the principle is that the cause must be as complex as the effect. In practice, what this means is that if you are looking at a change in embryonic development and believe that this developmental event is what caused an evolutionary event, the developmental event must be at least as complex as the evolutionary event.

Keeping this principle in mind, you can propose an idea, a hypothesis for how a cause led to an effect. And you can test it in the laboratory, by making a prediction. For instance, in the case of the bird hand, there has been a debate about how the five digits of early dinosaurs led to three digits in later dinosaurs and finally to what have appeared to be three different digits in birds.

When four-limbed animals first appeared, the evolution of the hand (and foot) was still in flux. An early tetrapod, *Acanthostega*, had seven digits on its hind limbs and eight on the "hand" of the forelimb. The number of fingers and toes was reduced, until the standard body plan of tetrapods specified five digits. Over the course of time some of these digits have been lost or become vestigial in different animals, but in embryonic development, as the hand grows, the beginning of the five digits can be seen. Changes in the course of development result in three-fingered hands, in some dinosaurs, and in birds, although in birds those fingers have elongated and changed shape to form wings.

In observing the development of embryos, the limb buds can be observed. You can watch how, in certain animals, they appear and then are lost during development. In birds, until recently, only four buds had been seen. And the identity of these digits, as established by the conventions of embryology, was a puzzle. The digits are numbered I–V in Roman numerals, going from thumb to pinkie.

The small theropod dinosaurs, like coelurosaurs, that gave rise to the birds had three-fingered hands, and the fingers have been numbered as digits I–III. Birds also have three-fingered hands, of a sort, although the bones in the digits are part of their wings. But they appear to have digits II–IV, according to the observations of embryologists. If birds descended from dinosaurs, this arrangement would not make sense. And some critics of the idea that birds are dinosaurs argued that despite the overwhelming evidence, the digit discrepancy showed that birds could not have descended from dinosaurs.

One piece of contrary evidence does not demolish a larger idea supported by a varied body of evidence from the fossil record. So even if the puzzle of the digits remained unsolved, the descent of birds from dinosaurs would still be the most convincing account of bird evolution. But the puzzle did call out for a solution. Hans and Günter Wagner, a colleague at Yale, worked on the digit problem both together and independently, coming up with an answer that not only solved the problem but demonstrated the way developmental and evolutionary events could be linked.

The essence of their approach was that there are separate stages in development of the chicken hand for which symmetrical events exist in evolution. In development the first is the appearance of the autopodial field, an area or zone of cells that are organizing themselves to create the beginnings of a hand. The second is the growth of digits, and the third the differentiation of the digits into distinct sizes and shapes.

In evolution Hans mapped comparable stages, tracing a symmetry between developmental and evolutionary events. The evolutionary event that he points to as parallel to the stage in development when cells organize to become a hand, is the appearance of fish called tetrapodomorphs. These were the fish that preceded the move to land. They had four fins that look like they were thinking of becoming limbs, so to speak.

The next developmental step is the growth of digits, and the parallel evolutionary event is the appearance among these sorts of fish of digitlike structures in a somewhat jumbled handlike paddle at the end of the fin.

But there is a third stage in development, when these growing digits acquire an identity, a characteristic structure. In evolution that stage occurred with the appearance of four-limbed creatures, intermediate between fish and amphibians. These are tetrapods, like *Acanthostega,* with seven digits on the hind limbs and eight on the front limbs. These digits were different in structure, so that you could distinguish one from the other.

Acanthostega may not have been able to walk well on land. Its limbs probably helped it to navigate shallows near the water's edge. *Ichthyostega* was another tetrapod, also a shallow-water creature that may have been able to walk on land. It had seven digits on its hind limbs, of identifiably different shapes. A recent discovery, *Tiktaalik,* a four-limbed fish that has many of the characteristics of later tetrapods, is sometimes called a "fishapod." Eventually, four-limbed creatures colonized the land and settled on five digits, and all the shapes we see today, including wings, hooves, and the hands of concert pianists evolved from the five-digit hands and feet of our lumbering ancestors.

In both the developmental and evolutionary stages, each step is built on the preceding one, just as with the growth and evolution of the feathers that Richard Prum worked on. Digits begin to grow before they take different shapes and sizes. Just because a digit starts growing in the spot where we might expect the first digit to be doesn't mean that it necessarily has to become the first digit. That may be the normal course of development, but it can be altered experimentally, and it could have been changed in the course of evolution.

Suppose the bud (anlage) of the second digit appears, but the sequence of HOX genes and sonic hedgehog and bone morphogenetic proteins that would normally turn it into the second digit are altered. Then that bud could turn into the first digit.

That would suggest that the developing bird embryo could start on the path to developing digits II, III, and IV, but end up with digits I, II, and III. Arguments over development can become so elaborate that they are hard to follow, but what this would mean, in brief, is that the evolutionary road from dinosaurs to birds would be cleared up. If, however, this change in development occurred, then a fifth bud ought to show up in development and ought to be in the right place. And a fifth bud had not been discovered until recently, when it was observed and reported by Hans and two other groups as well.

The other groups used different techniques, which were suggestive, but not as definitive as the work by Hans, which tracked the condensation of cells as they developed into buds, and then into digits. Hans and Günter Wagner then joined together to work on interpreting the evidence Hans had developed. They have argued cogently that in the developing chick embryo the growth that begins as anlagen II, III, and IV develop into digits I, II, and III.

This work, however, has importance far beyond the specific case of the bird's digits. The application of the experimental process to issues of macroevolution, the same process Richard Prum and others have used, marks a new and more rigorous way to understand the past. Paleontology has given us wonderful creatures, dug up from the past. It has provided the raw material for analysis and tracking of evolutionary change on a small and grand scale. It has not provided the mechanism, however. The mechanism of evolution, molecular-level changes in DNA and gene regulation, has been studied in the laboratory but has been restricted to small changes. Evolutionary developmental biology puts the two together, and the result for all of us is a more coherent and detailed understanding of how evolution proceeds.

The lab work can be pushed another step, one that Hans and a few other researchers have just begun to approach, and that is to create an atavism. We can try to change the course of development in ways that would turn back the evolutionary clock. Once we have established, in the case of feathers, or digits, a sequence of development, and have a hypothesis about how changes in this sequence occurred during evolution, we can test our hypothesis. We can intervene to make the sequence of development, at the molecular level, what we think it was before the evolutionary change. We can tweak the developmental instructions given the embryo to see if the ancestral state can be re-created.

A danger is that you could simply create an effect that looks something like the ancestral state, but you might have found another route to produce a superficially similar result and not have rewound evolution at all. There are ways to protect against such a result, but this a murky area and one little explored so far. Nonetheless, in principle, if you can turn back one evolutionary pathway, you ought to be able to turn back several. If you can do it for one trait, why not for several? Why not turn a chicken into a dinosaur?

Arguments can go on forever about evolution and whether it should be taught in the schools and about the abstract nature of science and evidence. But just as a picture is worth a thousand words, I thought a living dinosaur would be worth a thousand court cases in the visceral effect it would have on schoolchildren.

6

WAG THE BIRD

THE SHRINKING BACKBONE

Most species do their own evolving, making it up as they go along, which is the way Nature intended. And this is all very natural and organic and in tune with mysterious cycles of the cosmos, which believes that there's nothing like millions of years of really frustrating trial and error to give a species moral fiber and, in some cases, backbone.

—Terry Pratchett

Hans Larsson is a fast walker and a fast talker. You need to be fit if you want to keep up with him on the hills of the McGill University neighborhood in Montreal, let alone on the remote islands of the Canadian Arctic where he searches for fossils in summer fieldwork. He talks the way he walks, freely swinging in a fast-paced lope from the philosophy of science to genetic probes to the rich Cretaceous ecosystem he is exploring at another field site in Alberta.

Like many paleontologists he has been fixated on dinosaurs since childhood. He is, however, unusual in the breadth of his intellectual interests. Just as he seems impatient with a slow walking pace, he is impatient with the limitations of traditional paleontology. He is one of the scientists in the forefront of merging paleontology and molecular biology in an effort to connect major evolutionary changes—the development of new species and new characteristics, new shapes and structures, new kinds of animals—to changes in specific genes and their regulation.

He came to his current mix of research because he found himself unsatisfied with the business of collecting and categorizing fossils and drawing inferences about evolution from the fossil record.

"The reason that I was initially disenchanted with dinosaur paleontology is that these things were not testable. Just sort of stories and scenarios. Anybody and their mom and dog could come along and join the party. So it needs to be rigorous. And there needs to be some testing and developing these things across interdisciplinary approaches. And so including experimental embryology and ecological approaches to it, that's keeping me satisfied with it."

Anybody who wants to shake up traditional scientific approaches, bridge disciplines, ask new questions in new ways, is a researcher after my own heart. Using embryology to test ideas developed through paleontology seems to me to be a big part of the future of evolutionary biology. And Hans has been doing research right at the heart of the transition from dinosaur to bird. What is even more intriguing to me is that he is interested in pushing experimental embryology forward by reactivating dormant genes or changing the regulation of active genes to bring back ancestral traits that have been lost in evolution.

Another part of his approach to science, consistent with the desire to make paleontology more rigorous, was a concern with the philosophy of science, with the nature of proof and evidence and experiment. This kind of concern, rare among experimentalists and field paleontologists, is another aspect to his unwillingness to accept the status quo. He wants not only to make paleontology testable by laboratory experiment, he wants to define the nature of testability and what constitutes an experiment.

Over the forty years or so that I've been deeply involved in dinosaur research, what satisfies me the most has changed. First it was finding new fossils. Next it was changing paleontology by pushing it to bring new sorts of research techniques into practice. For the past few years it has been teaching. The greatest pleasure now is watching graduate students who are smarter than I am turn over old ideas and break new ground.

All teachers hope to pass on something to their students. For me, it's not specific knowledge or technical expertise. My graduate students quickly outstrip me in lab skills and knowledge of molecular biology. My goal is more like that of a high school science teacher who recently introduced his class to the theories of Georges Cuvier, a genius whose career straddled the eighteenth and nineteenth centuries. He was the first paleontologist, and proved the reality of extinction. He did not suggest that new species evolved, however. His theory was that the earth was incredibly old, and stayed largely the same, with periodic catastrophic changes, or revolutions, that caused extinctions. He did not see the reality of gradual geological change over time that molded the earth.

The teacher presented Cuvier's ideas to the class as solid, well-proved science. Some students disagreed, but he argued them down and, with a combination of his greater knowledge and his position as the teacher, eventually convinced all but one student. Once he had done so, he made a sudden about-face, now revealing to his class that Cuvier's ideas were, in fact, erroneous. He congratulated the student who had held on to his point of view, and, turning to the class again, warned them never to believe something just because a teacher said it was true. That's what good teachers, at any level of science, or any other field, for that matter, hope to pass on to their students.

Hans is not a student of mine. In fact, I could be a student of his in development and molecular biology. But he has exactly the kind of agile and restless mind that any teacher would look for in a student, and that any scientific discipline needs to keep it alive. He also has the invaluable ability to merge disciplines, to jump from fossil collecting and analysis, to laboratory experiment, to the philosophical examination of how science should proceed in its investigation of the past. A physics or chemistry experiment can be done in the laboratory and reproduced until the standard of scientific proof is clearly met.

The history of life is more elusive. We have the fossils. We have developmental biology. We have molecular biology. All are now being merged in the study of the history of life in evolutionary developmental biology. Along with some of his colleagues, like Günter Wagner of Yale, he has been concerned with establishing an accepted set of logical parameters for forming and testing hypotheses in evo-devo.

When a field is new and attractive, it is easy to make leaps beyond what we actually know. It is so clear in general that any evolutionary change in an animal must be a change in development that it is very tempting to start connecting the dots quickly, linking specific developmental and genetic changes to changes seen in the fossil record. But over the past two centuries the use of embryology to discover clues to evolution has faded in and out of favor. One reason is that for a scientific field to prosper there must be agreement on how to assess the evidence, and what logical steps lead to falsifying or supporting a hypothesis.

Collecting and cataloging fossil bones, the heart of vertebrate paleontology, has been primarily a historical enterprise, one of collecting information and looking for patterns in that information. You could test some conclusions, but experiment was not really part of the discipline. Instead you might conclude, on the basis of discoveries of dinosaur eggs and young in preserved well-drained highlands in Montana that dinosaurs preferred this kind of territory for nesting. You could predict that similar formations would produce young and eggs around the world and that wetter locations would not.

That is a self-serving example, since it was my prediction and more eggs and young were indeed found in those kinds of locations. Still, this is more like history than chemistry. And I've had my fair share of hypotheses that have been proved wrong.

Laboratory science, in particular the study of microevolution, has been conducted in a different fashion. You could suspect, say, that a particular growth factor is important in the formation of the tetrapod hand. So the hypothesis might be that if that gene were absent or nonfunctional, the hand would not develop. With mice, one can knock out a gene. You can engineer the mice so that the gene is absent or silenced and see what happens in development of the embryo. If the hand develops perfectly, you have falsified your hypothesis. If it does not, you have good evidence that the gene in question does what you thought it did.

With flies and worms such hypothesis and experiment is relatively straightforward. The genetic systems are simpler, the generations shorter. All sorts of evolutionary hypotheses can be tested. But these all have to do with small changes. What about a significant change in form in which something new is introduced that hasn't been seen before in evolution, something like the appearance of limbs, or hair, or feathers, or lactation?

Well, it can be investigated in the laboratory with intellectual rigor by adhering to the notion of symmetry that Hans used in tracking the evolution and development of the vertebrate hand and that Richard Prum identified in the evolution and development of feathers.

REWINDING EVOLUTION

But there is another way that it can be tested that has hardly been attempted, and that is to run the tape of evolution over again, to use our ability to intervene in the course of development in the chick embryo (or other embryos) to reverse evolution. This is a profound advance in the kind of experiment available to test evolutionary theory, and it depends entirely on the progress that has been made in evolutionary developmental biology. It is only because we can match developmental events to evolutionary events, only because we now have both the fossil record, which shows us the path that evolution has taken, and the developmental record in extraordinary detail, that we can link the two.

It is hard to overestimate the importance for understanding evolution that a detailed record of development offers us. We have the ability to map precise developmental pathways, not just in terms of the observed patterns of how cells organize and differentiate, but in terms of which genes are activated, and when, and which growth and signaling factors are present in different areas of the embryo, and when, and at what levels. We have the tools, with probes that tell us what proteins are directing growth and development, to discover and write down the entire program of growth for an organism. The complexity of this for a human being would be overwhelming. But in principle we could acquire that information, and the computing power to organize and understand the information that grows by leaps and bounds.

To be realistic, of course, we are not close to such an achievement. Look at the progress with *C. elegans,* the first multicellular organism to have its genome mapped. Researchers have also mapped its development cell by cell. That is to say that for the body cells (apart from the gonads) every cell in the worm's body can be traced from the fertilized egg to the fully formed worm. But we do not have the accompanying set of instructions, when each gene turns on and what concentration of each growth factor and signaling factor occurs at each location and at each stage in development. In fact, although the genome has been mapped, that does not mean that scientists know each gene. The sequence of every section of DNA is known, but it is another thing to fish out of that database which sequences are genes, and what their function is. That process itself is an enormous challenge. But it is conceivable that a truly complete instruction book for the development of *C. elegans,* from one cell to fully formed adult, can one day be compiled.

But a full instruction book is not necessary to try to rerun evolution. With birds, for instance, the absence of a tail, the difference between wings and grasping forearms, the absence of teeth, are all subtle evolutionary changes on a basic dinosaur plan. Perhaps if we imagine the dinosaur plan as a house plan. We might be going from a Cape Cod to a saltbox. Or you might think of the evolution of cars, since the way technology develops has some parallels to organic evolution.

Perhaps we could look at four-wheeled contrivances, like the simple cart, as similar to the first tetrapods on land, and the modern profusion of motorized vehicles as the modern world's profusion of reptiles, amphibians, and mammals. The birds, as dinosaurs, are included with the reptiles. Perhaps they are sports cars. This analogy clearly does not hold up if you look at it carefully. But the point is that trying to reverse-engineer a Model T from a Corvette is not as complicated as going all the way back to the invention of the wheel. Perhaps a closer analogy would be to changing the manufacturing process to leave out fuel injection and vary body shape. The chassis would stay pretty much the same.

Development is tougher than car manufacturing. Cars don't really evolve. They don't self-assemble starting with one part. And we have fully accessible and detailed manufacturing plans for cars with and without fuel injection. But if we have a detailed manufacturing plan for one part of chick development, the wing or tail, why couldn't we go run back the development process again, this time triggering the signals to produce a grasping forelimb or a long tail?

If we have proposed an evolutionary process like the one for feathers or for the change in digit identity, we could rerun that section of development, changing the developmental pathway as a means to testing the validity of our proposed evolutionary pathway. At last, we would have a truly experimental way of studying macroevolution.

So, why not grow a dinosaur? At least that's the thought that came to my mind. Leaping over the many details, it seemed so obvious that if fairly small changes in development, which adjusted the timing and concentrations of growth and signaling, could have led to the evolution of birds from nonavian dinosaurs, we could readjust those changes in development and get a dinosaur. Thus my idea of growing a dinosaur from a chicken embryo.

I talked to a variety of scientists about the project, some in independent laboratories in Asia, who were ready to jump right in. One of the limiting factors was money. To take on the whole project at once I would have needed millions. The National Science Foundation does not provide grants of that size to paleontology. And, as is often the case, I had come up with an idea that didn't quite fit into the standard categories of science. Millions to turn a chicken into a dinosaur might not be the most politically popular scientific grant. It was highly speculative, and likely to make people nervous about changing life in a fundamental way. Not to mention the issues of whether this was fair to the chicken or not.

I understand some of the concerns, although a number of them are, frankly more political than ethical. We do genetic engineering on mice all the time. By knocking out one gene or another we produce obese mice, diseased mice, crazy mice. None of these mice are going to survive outside of a laboratory environment, because most of the knockouts are disabling. So we don't need to worry about invasions of fat, bald, schizophrenic mice. But the ethical principles are the same. Because these knockout mice are so important for studying basic genetics and for understanding human disease, the societal consensus is that the research is worth doing. Not everyone agrees, of course. There is a large animal rights movement, and some people within that movement are opposed to all experimentation on animals. So far they are a minority of the general population.

If someone were to achieve complete success in growing a dinosaur, so that I could present the adult animal along with the scientific paper at a meeting, one question would be whether there was a risk of such animals escaping. Could this be Jurassic Park? Well, one animal could conceivably escape, but it would at best have the chances of survival that a lone chicken would have. It certainly could not reproduce in kind, because we are only talking about causing changes in the growth of the embryo, not changing its genes. So the apparent dinosaur would still have a chicken genome. If by some miracle it did mate with a hen or rooster, depending on its sex, the result would be an old-fashion chicken. If it died, we could stuff it and roast it. It would taste, as the proverb says, like chicken.

Another question might be about cruelty to the animal. I won't suggest that pretty much any life would be better than that of most of the billions of chickens that are eaten each year, at least not as a logical defense. But if we were able to do this correctly, we would have not a carnival freak, but a creature with a functioning tail and forelimbs, and useable teeth. If there were any indication that the chicken-dinosaur were in pain, we would not continue with the experiment. That would entail killing the chicken, but if we decide as a society that killing chickens is unethical, far more will have to change than us giving up an experiment. So, unless the chicken would suffer mental anguish at its dinosaur appearance that we could not detect, I would argue that the experiment would not be cruel to the chicken.

In any case, I did not try to convince Hans to try to grow a dinosaur because, although that is the goal I have in mind, and

I hope to find a way to push, pull, or cajole researchers into that direction, there are many, many small steps to take first. Each of them is difficult enough.

Hans was already researching how the tail in birds first got shorter and then disappeared over the course of evolution. I thought, well, why not look at it from the other direction? Suppose we were going to go backward, to try to bring back the tail, what would we do then? I gave about forty thousand dollars from my own pocket to pay for a postdoctoral researcher to work on this problem. Hans is continuing to pursue the research on his own now.

What attracted him to the project was the chance to push paleobiology a step further into rigorous laboratory testing, and to push evo-devo a step further by creating experimental atavisms. In development an atavism is an ancestral characteristic that appears in otherwise normal embryonic growth. For instance, human infants are sometimes, although rarely, born with tails.

That seems to be an atavism, but there is always a question whether the characteristic is simply a defect or mutant form that, in its structure, looks something like we imagine a tail would look. Or is it a trait that we still have the genetic information for reappearing because of a change in gene regulation prompted by the environment? The medical literature includes references both to true tails, with muscles, nerves, and blood supply, and pseudotails, which are simply something that looks like a tail. All of these are small—a few inches long—and clearly an accident of development, even if they have muscles and nerves. They are not the long tails we might imagine on the primate ancestors of chimpanzees or humans.

HEN'S TEETH

An experimental atavism is something we would produce deliberately. If you can intervene in development of the embryo to produce an ancestral characteristic, that would be an experimental atavism. A very few experimental attempts to achieve this kind of thing have been done. Chicken embryos have been induced to grow teeth. In one case, mouse tissue was transplanted and teeth were produced, so this was not a true atavism, these were not the teeth that we would have seen in an ancestral bird or dinosaur. What that experiment showed was that the tissue of the developing mandible in the chicken was capable of responding to the signals for tooth growth.

In another experiment, however, changes in the presence of growth factors produced teeth in a chicken without any transplant of tissue. The teeth were consistent with those of archosaurs, the group that includes birds, nonavian dinosaurs, and crocodilians. If that report is correct, then the researchers achieved a true experimental atavism.

"The idea," Hans said, "is that if there is an historical event, say an evolutionary transformation, that those events would not only follow some set of developmental rules, but some of those signatures of that particular developmental change or modification should be or may be present in the descendant forms." The signature would be in the molecular biology of development, a moment, for instance, in the course of an embryo's development when growth in one direction stopped, and restarted in a different way. The cause of this change could be found by testing for changes in the concentration of the different proteins (growth or signaling factors) that promote and direct development.

The disappearance of the tail seemed to be a good evolutionary moment that would have left a signature in embryonic development. Primitive birds, like *Archaeopteryx*, still had tails, but modern birds have lost them. It seems a good bet that this was a simple change that occurred, a turning off of the growth program that was keeping the tail going. Find the chemical switch, flip it the other way in embryonic development, and the result would be a bird with a tail.

Before he began an attempt to create an atavism, Hans had already done research on tail development and he had encountered several surprises. The fossil record showed that in birds long tails gave way to short tails, and then to no tails. It seemed that in the chick embryo one would find a short tail, with few vertebrae, beginning to grow, before it was stopped. But the tail was not a system that had been looked at closely in embryology. "There is no body of literature like there is on limb development for tail development," Hans said, "because it's a much more underappreciated structure and humans don't have long tails."

It's not that we are, as a species, incurious about other species. Look at our fascination with dinosaurs. But money and attention for embryology tend to flow toward research that may have an application to medicine. We are likely to pursue with more energy and money an understanding of a devastating birth defect in spinal development, for instance, than what happens with tail growth, although as it turns out the two may not be so far apart.

At the very beginning of his research on the tail, Hans found a couple of curious things. First, and this was not the big surprise, the chick embryos were not starting out with a short tail. Although the chicken and other modern birds have only five vertebrae plus the pygostyle, the embryo started out with the beginnings or buds (anlagen) of eighteen vertebrae. This is in an embryo about an inch in diameter, the size of a quarter. "So there's actually quite a long tail in these embryos," Hans said. The growth of a tail begins in the embryo and develops in a clear way, "adding on more vertebral anlagen on the end." But, he said, "Then I found that at a particular stage of development everything comes crashing to a halt."

The growth of the notochord, which preceded the spinal column in evolution and helps organize the growth of the spinal column in the development of all vertebrate embryos, was disrupted. Instead of continuing to grow from front-to-back, and guide the growth of the vertebral column, at a certain point the cells at the growing tip of the notochord "look as if they're disintegrating."

"It stops, becomes disorganized, and then makes a ninety-degree turn." This was the shocker. Tail growth is at least superficially similar to limb growth, and in limbs there is a group of cells at the growing tip that plays a major role in organizing and directing growth, and produces a lot of the proteins involved in promoting it as well.

"The tail seems to have something similar in birds," Hans said. And, "It was known to be present in mice and zebrafish." It was not present in another popular vertebrate, *Xenopus,* the African clawed frog, perhaps the ugliest of all lab animals, from a human point of view of course. "It seems to be just that *Xenopus,* being a frog, and from Mars, is doing something slightly different." Frogs, apparently, grow by their own rules.

So what Hans found was going on as the chick embryo grew was that the group of cells (the ventral ectodermal ridge) that was conducting growth just disintegrated, and tail growth stopped. Immediately after the end of tail construction in one area, however, what seemed to be a second, similar area of growth, another ridge, started nearby, and the notochord, the scaffold on which the spinal chord and tail are built, took a ninety-degree turn.

"It stops, becomes disorganized, and then makes a ninety-degree turn toward this new structure, which is *totally unheard of.*" No one had found that before. "It's at this point, when the notochord starts turning, that cells start condensing around the tip and start forming cartilage and then bone. This is the beginning of the pygostyle.

"This is something again unheard of. It has only been recorded in salmon, and salmon don't have a pygostyle, and they're not closely related to birds. So salmon and birds have done this very unusual trick of ossifying the end of the notochord. Nothing else does this. It might be a way to really stop tail development."

These findings threw a monkey wrench into the evolutionary tree, the phylogeny of birds and dinosaurs, because Phil

Currie, now at the University of Alberta, had described a dinosaur in 2000 that has what looks like a pygostyle. The dinosaur is an oviraptoroid, related to birds, but not directly ancestral. So either the pygostyle evolved twice, in different lineages, or the details of the dinosaur-bird lineage needed rethinking.

With this understanding of tail growth Hans and his postdoctoral assistant began in the winter of 2007 to try to make a chicken embryo's tail grow. This was the beginning, in my mind, of growing a dinosaur, although that's not what they were attempting. But it was the first attempt to modify the chick embryo in that direction.

They began snipping off the tip of a growing tail at one stage and stapling it, with fine tungsten wire, to a later stage, to see if the growth factors in action during early tail growth could override the stopping signals at stage 29. By convention scientists have broken down development into forty-six steps known as the Hamburger-Hamilton stages, after Drs. Hamburger and Hamilton, of course. The forty-six steps cover twenty-one days.

Presumably, a transplanted tip of a growing tail might direct an older tail to keep growing. If this work showed some effect, then the obvious candidates for signaling factors would be sonic hedgehog and fibroblast growth factors. The next step was then to add, before stage 29, retinoic acid, which is known to stimulate the release of sonic hedgehog, by injection or in microscopic beads. The hope was to keep growth going.

In either case, transplant or retinoic acid, if the tail were to keep growing, chemical probes could be injected to find out what genes were being expressed, what other growth factors were present. The probes are designed to find the messenger RNA that carries the instructions for the manufacture of growth and signaling factors. Messenger RNA is easier to find than the growth factors themselves, which are proteins. Because we know what the messenger RNA sequence is for sonic hedgehog, for instance, we can use a mirror-image stretch of RNA, called antisense RNA, that locks on to the right RNA stretches.

Hans and his postdoctoral assistant tried transplanting tips from stage 22–24 to stage 27–29, and then also from stage 17–19 to stage 22–24. They used up quite a few embryos, of course, but there is no other way to do these sorts of experiments. And in a world that eats eggs and chickens, the supply of fertilized chicken eggs seems almost infinite.

Transplanting did not cause any continuation in the growth of the tail. In fact, the tails stopped growing, perhaps because removing the tip was enough of a wound to affect growth through physical damage to the tail, or perhaps because there was a group of cells at the tip that formed the organizing center for tail growth and once that was gone, transplanting of another tip was not effective.

The use of retinoic acid did seem to work. The tail did continue growing and adding vertebrae, but not in a way that was meaningful to the experiment. The retinoic acid, Hans said, "pushed tail growth to the upper range of normal development. So it had some effect, but it didn't break it out of the cycle."

Even if a full tail had grown, however, it would not have been proof that Hans had been able to reactivate an ancestral pattern of development. It might have been a freak of sorts, because, he was realizing, he did not have a good enough baseline model of how the tail grows and turns into a pygostyle in normal development. It became clear to him that the tail was a far more complex system than he had imagined, and that if he were to get more than a hint of growth, he would need to know more about the normal development of the tail and pygostyle in the chick embryo. The initial hint of growth was encouraging. He was obviously on the right track, but he needed to know much, much more in order to avoid creating something that looked like a longer tail but was simply an embryologist's trick, finding a way to promote growth that was not an ancestral pathway.

Suppose, by throwing different growth factors at the tail at different times, a way could be found to grow a full-length tail. Unless he knew the normal pathway of development in detail, and could say exactly at which step he had intervened, and why this was likely to be the evolutionary change that resulted in no tail, all he would have achieved would be a circus attraction. As he put it, "One danger would be that we're just sort of making a new anatomy. But I want to be sure that we're playing with or manipulating the exact same system that is there normally."

The pitfalls were clear. To take an extreme example, thalidomide given to pregnant women can result in babies being born with short flipper- or finlike limbs. But if these were to be claimed as atavisms—throwbacks to an earlier ancestor—the claim would be ludicrous. The same foolishness might misguide us into thinking we were looking at reverse evolution in the case of the tail. What we are hunting for is a change in the same system that changed during evolution.

Had the shotgun approach worked and a tail grown out to all eighteen vertebrae, Hans could have gone back to use the genetics and chemistry of that change as a starting point for digging down to the basic developmental genetics of tail growth. Or, if he were working on the limb, turning the wing into a forelimb, he could have gone to the scientific literature, which is wide and deep on limb formation. He would have found there enough work to create a baseline system of gene activation during limb growth and to know whether a change in limb growth was a modification of the baseline system that might be related to how it evolved, or whether it was simply a crude way of throwing off the course of development.

THE UNKNOWN TAIL

With the tail, he found, there was no such foundation of basic research describing how it grew. Research had been done on tail initiation but not on how its continued growth was mediated, and by what genes. He found he had to start from scratch and do basic developmental research. "We're essentially having to try and map out these very basic genes and very basic research that has to be done just to make some sense of normal tail development."

The tail is not a major identifiable part of some of the most studied evolutionary transitions—from fish to land, for example. Limbs have gotten quite a lot of attention. But, says Hans, tail growth is actually more complex than limb growth and potentially more profoundly important to development.

"The tail is, in essence, a little more complex than the wing or the leg because it has a lot more structures to it. The limb is just a bag of cells growing out distally and it brings with it the skeletal structure, the muscles, the nerves and blood. But that's about it.

"The tail has all of that, plus it has two other things that are completely different from the limbs. It has the central nervous system in it, which is crucial. Secondly, it has the notochord. And the notochord is something that's very, very basal within all chordates. All chordates have the notochord, and it's another signaling center. The notochord, the central nervous system, and the surrounding mesoderm all set up a sort of cycling of events that helps to generate front-to-back segmentation and front-to-back polarity."

This is about as fundamental a process of growth as there is, determining bow and stern. The whole tail system and its development may turn out to have major import in developmental biology. "Clinically, central nervous system, notochord, and axial patterning are really fundamental for developing animals. There are more defects known at that level of development than there are at pretty much any other sort of developmental hierarchy."

What has happened is that the questions we asked have led Hans to take on a fundamental aspect of development that has not been studied in any detail. "We're helping to reveal a whole new developmental system, which might be very fundamental because it involves the whole body axis, the front-to-back axis, and making it longer and shorter, and so the mechanism driving this or controlling that level of development could have very, very profound implications."

Why, for instance, do we have so many vertebrae? As with so many other aspects of animal design, there is clearly no design going on. Instead we see changes built upon existing systems, and evolution can only be understood by seeing previous systems and finding out how they evolved.

If the basics of tail development can be understood, then it becomes possible to ask these questions, make hypotheses, and test them in the lab by manipulating the normal program of development. Hans is trying to get to the most basic level of gene regulation. Vertebrae are derived from body segments called somites. There has been a lot of study and a body of knowledge built up about how each somite is built and its boundaries defined. What Hans hopes to find out is what turns on the somite generation engine.

If that key can be found through mapping out this complete developmental system, then the work can turn back to evolutionary biology. If the key can be found that turns on and off the tail growth process down at the somite level, the genes can be sequenced, the process mapped out, and he will, he hopes, "find out exactly what's changing on a molecular basis across very different-length animals—alligators and birds, for example."

This would be a major breakthrough toward understanding the molecular basis of large-scale evolutionary change, "whole skeletal level changes across millions of years." Nothing like this has been done yet.

Of course, this began with a paleontological question about birds and dinosaurs, based on fossils. And it is an interesting lesson from the point of view of interdisciplinary studies. A paleontologist says, why not mess with a chicken embryo so that it grows into a nonavian dinosaur? It can't be that hard. After all, they are both dinosaurs. They share a very similar skeleton. In the grand scheme these are small adjustments to a basic body plan and such adjustments, we are assured by evo-devo, are the result of changes in gene regulation, not a complete new suite of genes.

So let's see what regulates limb growth, the growth of teeth, and tail growth—that ought to be the simplest of all. Hans tackles the question, assuming at first that it should not be that difficult, and lo and behold, a hole in our knowledge of vertebrate development becomes glaringly obvious. We don't really know, at a molecular level, what runs the growth of a tail, which, it would seem, is central enough that it could be said to wag vertebrate development.

I know Hans agrees, and I certainly think, along with other scientists who are pushing for interdisciplinary studies, that beyond the obvious value of molecular biology to paleontology, there is the value of having the fossil specialists work hand-in-hand with lab researchers. It is the fossil record that gives us the story of how life evolved and raises questions that can be pursued in the lab. The tradition of molecular biology is to look at the smallest changes in the greatest detail. The fossil hunters are the ones who understand the grand sweep of evolution. And it is good to think of us coming back into the fold of biology after having been given the cold shoulder for decades.

Still, it is a stumbling block and a delay in the plan to grow a dinosaur.

What Hans is doing now is developing a baseline model of how the tail grows. This will involve labeling cells in the growing embryo tail, using microinjections of dye to follow the pathways that cells take as the tail develops into the pygostyle. He needs to see where zones of growth and organization move as the embryo grows. And he needs to test for what is going on biochemically as growth occurs. So his students will test for concentrations of sonic hedgehog in the growing tail every four to six hours during normal tail growth.

They will also be testing the presence and activity of another family of proteins, the Wnt family, which includes Wingless (Wg) a famous protein and gene (the genes and the proteins they code for share the same name). The Wnt proteins are signaling factors that are involved in controlling patterns of growth. They were first identified in fruit flies but, like many of the most important factors, are found to be significant in all sorts of animals. In vertebrates, the Wnt pathway, as it is called, is active in limb growth and all sorts of other areas.

A great help to Hans in his attempt to identify which genes are active at different points in tail development is that, although the tail itself has perhaps not gotten the attention it deserves, the development of the chicken embryo has been more than thoroughly investigated. Researchers who study other aspects of chick embryology do so by staining the embryo at different points in its growth to show the activity of the genes they are looking for. They may be interested in the skull, ribs, forelimb, or hind leg, but chick embryos are small, and there is no point in trying to restrict their tests to one tiny area. So when they use a stain to show gene activity, the stain affects the whole embryo.

The limb investigators do not look at the stain results in the tail, but they do preserve the images of the chick embryos. So Hans and his students combed through the literature to find indications of what genes were active in tail growth. Having identified the target genes and their protein products they will record in more detail their activity in sections of the growing tail.

Once a record is compiled for the degree of activity and the three-dimensional location of activity of sonic hedgehog and the Wnt family of genes at 104 to 156 stages in the growth of the tail (the work will be done by a postdoc and several independent-study undergraduate researchers), all the images will be fed into a computer and students will work to create a digital 3-D model of the growing tail.

"Then we can actually point to a few of those genes and say, that's the one that we really want to hit aggressively."

At that point, this first step in the attempt to grow a dinosaur will really begin, that is, the step of finding the gene or genes that stop regular tail development and create the pygostyle. It may seem ironic, given the complications, that Hans picked the tail as the easier system in which to create an atavism, easier than the wing. But the fact is that although the growth of the tail is very complicated, the action that turns that growth off may be quite simple, whereas there would be many detailed moments of turning on and off different genes to cause forelimb rather than wing growth.

Hans compares the situation to a mechanical one. "It's kind of like the key to a car. You could turn the key on and the motor will run and produce all these patterns and rhythms coming out of it. Once you turn the key off, then it stops.

"And the key is relatively simple, compared to the rest of the car. I think that's the kind of system we're dealing with. Or I'm hoping."

As for the dinosaur, even within the egg, Hans says, "The experiment I'm sort of envisioning at night is that you have a single embryo developing in the egg with multiple injection sites and multiple kinds of molecules, proteins, or morphelinos or things like this to be really fine-tuning the regulation of genes." He continues, "So we'll be able to inject different parts of the embryo at different times of development with different things. If we do that, if the timing and position are correct, we should be able to manipulate lots of different kinds of morphologies— feathers, wings, teeth, tails.

"It would just take a little bit of time to work out each one of those systems at very great detail, which we're now doing for the tail. And other people are doing for the limbs for clinical work. And teeth are being worked out by other people for mammals and such, and then if we can just sit down and play with all these in concert, which has never been done before."

The goal, in the end, would be to steer the embryo down the path it would have gone if it were something like a very early coelurosaur. If the classic genes in the chick embryo that have the codes for the proteins that are so essential for life and growth are very close to those of an ancestral, nonavian dinosaur, and if the changes, over more than 150 million years, have been almost all in regulation of the classic genes, then we could find the old pattern of regulation.

Metaphors can be dangerous, but we might imagine the development of an embryo as a story, a series of events in which every event determines what other events are possible. Just as in a novel, or the story of a life, every event has a consequence for the rest of the story. In life we don't know what the consequences of choices will be. Will a date lead to marriage? Will a job work out? Will a chicken salad sandwich cause food poisoning? With the embryo we can do the story over and over with different choices in gene regulation, rewinding the course of development. We don't have to give the embryo new genes, just adjust the growth factors and other chemicals that direct development. And by doing that we can see what must have changed during evolution, and what the old pattern of regulation was.

If we learn enough, this will give us enormous insight into the fundamentals of biology, development, and evolution. It will also be the first step in growing a dinosaur.

7

REVERSE EVOLUTION

EXPERIMENTING WITH EXTINCTION

Way over 90 percent of all the species that have ever lived on this planet—ever lived—are gone. They're extinct. We didn't kill them all. They just disappeared. That's what nature does. They disappear these days at the rate of twenty-five a day. And I mean regardless of our behavior. Irrespective of how we act on this planet, twenty-five species that are here today will be gone tomorrow. Let them go gracefully. Leave nature alone.

—George Carlin

I hope that by now I have convinced you that we can start with a chicken embryo and hatch out something that looks like a nonavian dinosaur. Will it really be an extinct animal brought back to life? Would a tail, claws, and teeth be enough to say that we have brought such an animal back from extinction?

No. We would have brought back some of the characteristics of the dinosaurs. We would have used the signatures left by evolution in the chicken's DNA to rewind evolution. But we would have re-created ancestral traits, not the ancestor itself. We could never truly re-create a species or genus that was lost, unless we had a complete genome. If we reached that point, then there would be philosophical disputes about whether we could really bring back an animal from deep time. There would always be gaps in our knowledge and, not ever having seen the animal, we could not be absolutely sure. But we would have brought back an animal that, by all our tests, was identical to the extinct animal.

This seems unlikely for animals from deep time, for practical reasons in retrieving a full picture of an ancient animal's genome, as well as other influences on its development. But the more we pursue creating experimental atavisms, the closer we will come to this achievement, and the more we will learn about evolution, which is the deeper goal. The fundamental reason for attempting to rewind evolution is to learn how evolution occurs. Like a teenager with an old car, you take it apart to learn how it works.

On a shorter time scale we could certainly come close enough to re-creating an actual extinct species, but the achievement could well be an empty one. Let's use the ivory-billed woodpecker as an example. This was the largest woodpecker in the United States and Canada and it is generally thought to have gone extinct perhaps half a century ago, although there have been many claims of sightings, including one that was published in *Science* on June 3, 2005.

Earlier that spring, when the announcement of the sighting was made and the paper on the woodpecker released to the press, I was planning this book with my coauthor, Jim Gorman, and he was called away in the middle of our conversations to fly to Alabama and write about the woodpecker. Since then, the sighting reported in the *Science* paper has been roundly criticized as a sighting of a pileated woodpecker, and no hard evidence in the form of a clear photo or video, or DNA, has emerged to prove the ivory bill's continued existence.

There are woodpeckers that are in the same genus as the ivory bill, *Campephilus*, such as the Magellanic woodpecker from South America, *Campephilus magellanicus*. The Magellanic woodpecker is not descended from the ivory bill, but they share a recent common ancestor, recent in evolutionary terms. It seems possible that if we carefully documented the embryonic development of *Campephilus*, we might discover a way to make a *magellanicus* embryo develop to be indistinguishable from an ivory bill.

Whether we would want to is another question. I don't know what there is to be learned in this case, since the differences are subtle, and the re-created ivory bill would not breed true. We could also sequence the genome of the ivory bill from skins saved in museums and compare it to *magellanicus* and find differences in the genome. If we found obvious differences, we could perhaps change the genome.

But even if we were able to create a bird that was indistinguishable from the old ivory bill, it would always seem ersatz, particularly since the dream of finding the ivory bill still living is about proving to ourselves that we were able to stop our-

The ivory-billed woodpecker is thought to have become extinct, although there are reports of sightings. Similar woodpeckers might provide a genetic basis for reconstructing the species.

selves from driving a beautiful species into extinction. And for science and conservation and our own sense of the planet, the issue is not so much the bird itself but the bottomland hardwood forests it lived in. Without them, re-creating the bird would be like re-creating tigers without a jungle.

Other experiments might also be undertaken with a good chance of success. We could probably make the embryo of a domestic chicken grow into *Gallus gallus,* the wild chicken that is its ancestor. And if wolves disappeared we would have a great reservoir of genes in the domestic dog. Dealing with mammalian embryos is, however, quite difficult. Still, we have the dog genome, we can get the wolf genome now, while they are still living.

These would be very, very small triumphs of reverse engineering. Even to an untrained eye a Siberian husky and a wolf are not so far apart in appearance and behavior. We would be reversing microevolutionary changes. An untrained eye might not see the difference in the first place between one woodpecker and another. And behavior would be hard to re-create since it would depend, no doubt, on environment as well as genetic heritage.

In contrast to those experiments, however, the one that I am proposing—or campaigning for, I suppose you would say—promises significant benefits both in terms of basic research and applications. Turning the clock back from chickens to dinosaurs would open up to us a method to tackle the major changes of macroevolution and help us tie them to changes in the control of genes. And what we find out about intervening in embryonic development, particularly involving the growth of the spinal cord, could prove of great practical, medical use.

In the attempt to re-create a dinosaur, we can't pick a species. That's too fine a target. In another way that distance and time are connected, the farther away a target is at a shooting range, the larger the area you need to aim for. The farther back in time you go, the larger the target you might be able to hit. Research would have to aim at something phylogenetically larger, perhaps at the level of genus, or family. The farther we go back in time the less information we have about the extinct animals. Once we are in deep time, we are dealing with animals that we have imagined based on limited information. In some cases species have been named on the basis of a tooth, or not much more. Even with a relatively complete skeleton, there are so many areas where we would have to guess. Wait, let me rephrase that. We would have to hypothesize, based on the evolutionary context of the animal and other information. The color of the skin? The way the animal moved? Evidence based hypotheses, otherwise known as educated guesses.

What we can aim at with some certainty are the characteristics that we know from fossils—size and skeletal structure, teeth, musculature, and in some cases skin. We can make reasonable conclusions about movement and diet, and good guesses on certain aspects of behavior. For some behaviors, however, we would need herds of dinosaurs, complete with the appropriate predators and environment, to observe them. In other words—Jurassic Park. That is not something I will see in my lifetime. And probably not something worth pursuing.

How much we will eventually be able to achieve is impossible to say. We can keep pushing the boundaries of knowledge and ability, which will continue to grow. But the scientific capability must be balanced against mundane questions of money and usefulness and profound issues of ethics and social responsibility.

I have no doubt that we can and will do what I've proposed, to bring back teeth, tail, and forearms with claws. It won't be easy and the money may not be forthcoming, but it will happen, and I'm convinced that it will be worth doing. I also think we could change the kind of feathers a chicken grows to make them more primitive. I think we could achieve a suite of changes in one embryo so that the resulting animal could hatch and live out a normal life span, eating, moving, and functioning without difficulty.

Beyond that, we'll have to wait and see. As embryologists work out the details of the program for tetrapod development, I would expect many barriers to reengineering extinct life-forms to fall. Right now, we could change limb growth with many small interventions—adding and suppressing growth factors in different locations at different times. We may, however, find higher-level signals that lead to a cascade of developmental changes so that instructions for forelimb growth, for example, do not have to be adjusted piecemeal. There may be fewer changes needed than we now imagine to prompt development of a hand rather than a wing.

We may never know the physiology of long extinct animals. We can intuit certain internal arrangements from living birds, but I doubt that we will ever know exactly what the inside of *T. rex* looked like. Still, the more we try to rewind and replay the tape of evolution, the more we will learn about how animals are put together, how they grow. And we have clues in the digestive systems, to take one example, of modern birds with different diets. Combining the variations found in living birds and what we can learn of the diets of extinct animals, we may refine our ideas about the digestion of an extinct animal. If *T. rex* was a scavenger, for instance, we might look to see how modern avian scavengers cope with their diet.

SHOULD WE DO IT?

None of these potential increases in our ability to reverse evolution, to be more confident of the accuracy of our ventures in developmental time travel, answer the political and moral question of whether this is the sort of thing we should do. To consider this question, or set of questions, we must think about benefits and risks, and about what the challenges might be to the ethics of such an experiment.

Among the potential benefits of causing a chicken embryo to develop dinosaurian characteristics is that this is a project that could capture the popular imagination. It could be a demonstration of evolution that would be felt at gut level by non-scientists who might be uninterested in the details of genomes and embryos.

Anything that brings home to the public the reality of evolution, and its place as the foundation idea of modern biology, is important. Anything that dispels the fog of confusion about science and religion would be enormously positive. I teach a course at Montana State University called "Origins." This is not a paleontology class for students who are specializing in studying some form of evolution. This class is taught by a theologian and a cosmologist as well.

We have students planning a career in science. We have those who are not. Some students don't express religious views, others make clear they are Christians, but not biblical literalists. Sometimes we have had young earth creationists. My approach to teaching the course is the same as my approach to teaching science in general.

I am not teaching or promoting or asking anyone to memorize and spit back at me the accepted understanding of evolution. What I am hoping for, always, is to get across the idea that science is a way of thinking that has no necessary conflict with religious or spiritual thinking.

The difference between science and religion or philosophy is sometimes said to be the kind of question asked. It's more useful, I think, to look at the kind of answers that are proposed to any given question. In science you come up with an answer that can be proven wrong—a falsifiable hypothesis. Why does a stone thrown into the air fall to the ground? is a question to which there is an answer—many answers, actually. The testable answers, the ones that can be proven wrong, are scientific ones. If you propose a physical force called gravity that works in a certain way, that's an answer that can be tested.

If I proposed that all things are drawn to the earth, that might seem a pretty good explanation until we started thinking about the moon and the sun and the stars and why they weren't falling to the earth. In any case, it's an answer that can be proven wrong, or incomplete. If I say that the stone falls because its spirit wishes to return to its home, that's a spiritual answer. I can't think of a way to test that to falsify it. So that answer is outside the realm of science. Theologians and philosophers may have a way to address it, but I don't.

One of the benefits of actually hatching a dino-chicken would be that it would be shockingly vivid evidence of the reality of evolution—not a thought experiment but an Oprah-ready show-and-tell exhibit. The creature would be its own sound- and vision-bite. It certainly wouldn't convince anybody who didn't want to be convinced. But it would cause discussion and thought. What I like about the idea of using a chicken that developed into a dinosaur as evidence of the reality of evolution is that it is more than an idea. It is an experimental result. And it calls out for questions. What is it? How did you do it? Is it a circus freak or a trick? What does it mean? Without staking out a position or starting a war of words, the animal would prompt a discussion that would have to end up with the mechanisms of evolution and its footprint in the genes of living animals. Even more than a fossil, it would cry out for explanation.

Creating a demonstration suitable for soundbite television is not, however a reason to do scientific experiments. In order to get to the point where the question—How did you do that?—could be answered, we would have to learn a great deal. And we would tie molecular biology to macroevolution. We would zero in on a significant passage in vertebrate evolution, the transition from nonavian dinosaurs to birds, and pin it down to molecular changes in embryonic cells.

The scientific significance of such a demonstration would be great. If we know the pattern of the release of growth and signaling factors, then we can read back to what genes are being turned on and off, and we can provide a flow chart for the growth of a dinosaur's tail and the changes that make for the growth of a pygostyle. The science would not be in the showmanship of presenting a dinosaurlike chicken. The more characteristics like this we can pin down, the more detail we have on the small molecular changes that cause macroevolutionary changes in shape.

This is the heart of the promise of evolutionary developmental biology, to show how molecular changes effect large and obvious changes in animal shape of the sort that we have always tracked in paleontology. We can see dinosaurs appear in the fossil record, radiate into an astonishing variety of shapes and sizes and behaviors over 140 million years, and largely disappear, leaving one strain of descendants, the birds. Experimenting with embryonic development in an attempt to reverse evolutionary changes and bring back atavistic characteristics is the logical next step, the way to use the knowledge of evo-devo to test hypotheses in the lab. Call it *revo-devo*.

Basic research is at the heart of science. And I don't want to underestimate the value of the search for knowledge for its own sake. Nobody goes into the study of dinosaurs whose primary goal is to seek breakthroughs in practical knowledge. Nonetheless, this is part of science, and society at large rightly asks what benefits will come from research. That is particularly important if public money is being spent on research, but it is also important if the kind of research makes some people uncomfortable, which is certainly the case when it comes to intervening in embryonic development, even of chickens.

Vertebrate paleontology may seem to be so remote from the daily problems of the modern world that it exists apart from society. It offers great entertainment and wonder—witness the popularity of museums' dinosaur halls. But, if I were to be harsh, to ignore the unpredictable ways new knowledge of all sorts can add to our lives, I might ask, What good is it?

There is an aspect of vertebrate paleontology that is highly useful and of great importance to us as vertebrates. That vertebrate body plan that is so resilient, that has survived and prospered through flood, volcanoes, mountains rising and falling, and the bombardment of the earth by comets and asteroids, is one we share with dinosaurs, chickens, and countless other creatures. And the most basic aspects of how a vertebrate embryo grows are ones we have in common with all other vertebrates. So it should not be a surprise that there are paleontologists who teach anatomy in medical schools, or perhaps anatomists who also have a passion for fossils.

The result of this commonality of life, in this case in the specific fraternity of four-limbed vertebrates, is that lessons we learn about the growth of any tetrapod embryos may have significance for the growth of human embryos. Hans's research on the tail, for instance, has led him to work on the spinal chord and notochord and to investigate how that growth can be disturbed or redirected. That can clearly be of use to medicine, since spinal cord defects are among the most common and devastating.

If we learn about the growth factors that signal the neural tube to continue developing, it's possible that this knowledge could be useful in preventing birth defects. Humans do not have tails. But we do have spinal cords, and the growth and development of the two are intimately connected.

In spina bifida, for instance, incomplete development of the spinal cord can leave an infant with painful and sometimes lethal birth defects. In the 1980s researchers pinned down the importance of folic acid to the development of the spinal cord in human embryos. This discovery was made partly by gathering information about the diets of pregnant women and the incidence of spinal-cord birth defects like spina bifida, and partly with animal research. The simple remedy of adding folic acid to the diet of pregnant women now prevents countless cases of these defects. Hans is pursuing basic research on embryonic development, but at such a fundamental level that it is likely to have applications far beyond chickens and dinosaurs.

The chances for reasonable success in building a dinosaur are very good, and the benefits for basic and applied science that may accrue from the research, whatever the end result, are potentially very large. And here I should mention again that the end result, a dinosaur, is not really the goal of the research. It is a target, a means to an end. The ultimate goal, and the end toward which the research is aimed, is to increase our understanding of evolution and development and the connections between them. We learn as much from mistakes as we do from successes. So, for instance, Hans's findings about the number of vertebrae in the chick tail as it begins to grow and the way in which the program of tail growth is disrupted and then redirected, are results well worth the work he has done. If he or someone else eventually manages to re-create a dinosaur, fantastic. If not, I am confident that what we learn along the way will be worth the effort.

Knowing that there are great potential benefits to be had in basic and applied science from trying to make a dinosaur, answers some significant questions about whether the research should be done. But there are other questions. Is it a morally justifiable act to play with life in order to go back in time? Is it cruel to the experimental subject? Is it dangerous to us or our environment?

I am not really going to answer all these questions, because morality and ethics are individual matters. I'm just not comfortable coming out with a statement that this or that practice is right, and the other is wrong. What I can do is to put the experiment I'm suggesting in the context of generally accepted scientific practice and common sense. I think that the experiment can advance science but that it does not really pose new ethical challenges. It is well within the kind of research now accepted in science. The deep into the waters of animal rights and the fundamental opposition to all experimentation on animals are subjects for another book, or several books.

Those are questions that apply to entire fields of science, to farming, to using animals for food, to keeping pets, to having zoos, to the ethics of farming, land use, population growth, to the question of what constitutes a person with legal rights and protections. They are worth discussing, but they are not questions I have answers to, or in which I have any particular expertise.

What I want to say is that the attempt to make a dinosaur as I'm suggesting fits within the common practices of science and medical research. On the big questions, it does not occupy such a special position that it needs to be discussed separately. It may seem extreme, but I don't think it is.

First off, what Hans is doing so far, and the only work that he is planning, involves working with embryos, none of which will hatch. Experimentation on an embryo, at his university and most universities, does not come under the rules that govern the welfare of animals involved in experiments. There are those who object to any such experimentation regardless of how it is performed or what the benefits are, but that's a bigger moral, philosophical discussion.

Experimentation of all sorts on chicken embryos is widely accepted and, I think, the correct assumption is that we are not causing the embryo pain. As to ultimately sacrificing the embryo, or a fully grown chicken, there are far greater injustices and indignities that billions of chickens face every day. Common sense would suggest that not allowing an egg to hatch, or humanely killing even a full-grown chicken, are actions that society recognizes as legitimate, given even the small return of a meal. The potential return is much greater here.

No one is ready to let an embryo experiment hatch yet. But when that point is reached, when the plan is to have a fully formed dinosaurlike chick hatch, then the experiment will come under review boards that deal with animal welfare. My sense is that providing a chicken with arms with claws instead of wings, with teeth, and with a tail, would not be cruel. In fact, if the atavistic structures grew improperly or were malformed in a way that would cause the animal pain, that in itself would mark a clear failure, since the whole point is to re-create functioning atavistic characteristics, not monstrosities.

There are research programs that do create monstrous animals of a sort. In order to understand human diseases, in particular the genetics of disease, many genetically altered mouse strains have been created by knocking out or inactivating a particular gene or set of genes. Some of these so-called knockout mice are obese, diabetic, hairless, or the mouse equivalent of schizophrenic. Others are prone to develop cancers or other inherited diseases. Clearly, and this is the case with other experiments as well, we are willing to cause suffering to these mice if it is a necessary part of a valid experiment.

Although there are plenty of examples of human beings acting cruelly, we also recognize cruelty when we see it, in a common-sense way. The committees that review animal experiments at universities have a variety of technical rules, but in essence they are designed to eliminate unnecessary pain, and to judge the value of an experiment if it will cause suffering to an animal. As a society we have, in the past, been willing to experiment even on close relatives, like chimpanzees, if the potential benefit were important enough, combating AIDS, for instance. That is changing now, but these are really decisions for the society at large. In the case of an adult chicken that had developed as a theropod dinosaur, the experiment would only be successful if the animal were comfortable and well-functioning.

That may not be so easy, because it may be necessary to make other changes in bone or muscle structure, or neck length, to allow a long tail to be functional. As for the claws and teeth, the chicken/dino would need to be able to use its forelimbs. They would have to have grown complete with proper nerve and muscle growth properly mapped to the controlling brain. The teeth would also have to be functional, and not interfere with the chicken's diet. The red jungle fowl, the ancestor of all domestic chickens, which can still inter-breed with domestic chickens, eats mostly insects, seeds, and invertebrates. Domestic chickens will eat almost anything including worms, salad, fruits, grains. As long as the teeth didn't

interfere with eating standard modern chicken feed, the creature would be okay. And that diet could be supplemented with
other foods like worms and crickets, for which the teeth might
be helpful.

IS IT DANGEROUS?

There is a whole range of possible objections that have nothing
to do with the health or life conditions of what we could probably call chickenosaurus. And that is fear for the environment,
for interfering with the delicate ecological balance of the planet.
Many people fear genetically modified crops, for instance, or
genetically modified foods. It seems to me that the odds of
harm occurring from eating genetically modified foods are
very small. There is, of course, always a small chance that something new will cause an allergic reaction in some people. Other
than that, the nutritional value of the corn or meat seems the
same. Genetic modification also occurs in traditional selective
breeding, or the kind of grafting and hybridization that goes
on in developing new plant and seed varieties.

Selective breeding does not, of course, move genes from
one species to another, and again there is some possibility for
surprise. But I am getting away from chickenosaurus. If the
embryo is not allowed to hatch, then it won't be out in the environment at all. If it were allowed to hatch, and somehow escaped, the only problem would be the chickenosaur figuring
out how to survive. It would not be a danger to the environment or to the billions of chickens in the world, because, as I've

described, we would not be changing its genetic makeup. By manipulating growth signaling factors we would be switching genes on and off at different times during development, but not changing the genes themselves. Genetically, chickenosaurus would still be a domestic chicken. And if it were somehow to breed with a chicken, the result would only be more chickens.

Think of the difference in height between some immigrant groups and their children. Better nutrition during pregnancy and childhood leads to an increase in height. But there has been no genetic change. The growing child has simply had more fuel and, with the same genes as shorter parents, has grown taller.

A less happy but similarly instructive example could be babies born addicted to heroin or cocaine because of a mother's habits. When the baby grows up there are no new addiction genes that it can pass on. So chickenosaurus would be harmless.

If our understanding of embryology and evolution reaches the point at which we know how to alter DNA to change the growth program, then we could make animals that would pass on their characteristics to their offspring. That will bring up another set of potential problems. It seems unlikely that chickenosaurs would take over an environment rife with raccoons, opossums, cats, dogs, coyotes, foxes, fishes, snakes, rats, and people. Still, altering the genes of an animal, as is done with knockout mice, is not something I am suggesting.

That would really produce the possibility of Jurassic Park, and attendant problems, although probably not vicious raptors rampaging through the kitchens of Southern California. More important would be the question of whether such animals would be functional in the outside world if they escaped. Even as an invasive species, they could disturb environmental equilibrium.

Bringing back an extinct animal, or a reasonable facsimile of an extinct animal, is one thing. Actually bringing back the full species, or one like it, capable of reproducing and spreading in the wild, would be something else altogether. George Carlin, the brilliant comedian and social commentator who died recently, had a great routine in which he talked about endangered species and extinction. His point, made much better than I can, was that saving endangered species was just one more example of human arrogance, of interfering with nature. He noted that well over 90 percent of all species that have ever existed are gone. "They disappeared," he said. "Let them go."

Finally, there is the question of where this research will take us. Having learned to redirect embryonic growth in chickens, we might well extend these abilities. Chickens are, of course, so much easier than mammals to work on because the embryos are large and encased in convenient containers. They can be kept incubated and growing on their own while we add beads or inject retinoic acid.

But such techniques can be refined, and the more they are successfully used the more we will want to use them. We could use what we learn to alleviate human suffering in children now born with neural tube defects. We might be able at some time to intervene during the growth of an embryo, to override problems and ensure that growth continues as planned.

On the other hand, there is public disagreement about what a defect is. Different lines of research have supported an idea that sexual orientation has a genetic determinant or is affected by maternal hormones during pregnancy, or both. These are hypotheses that have some evidence, not well supported theories. But if, in fact, sexual orientation in some mammals is shown to be largely or partly the result of the presence of a hormone during pregnancy, some parents might be tempted to test hormone levels during pregnancy and adjust them to influence the sexual orientation of their child.

Other interventions could be tempting, too, perhaps for improved intelligence or athletic ability. This would not be genetic engineering. Ethically, someone might argue that it is no different than giving folic acid or intervening in some other way to prevent neural tube defects. But there is general societal agreement about the desire for babies that are healthy and not in pain, while there is bitter disagreement about other interventions. Would we want a world where some mothers receive treatment during pregnancy to increase the intelligence of their children, or to prevent them from being gay, or to cause them to be gay?

These may seem farfetched, but genetic testing of embryos is now in use with in vitro fertilization, mostly to screen embryos for known genetic diseases with devastating effects like Tay-Sachs, which can cause profound mental and physical disabilities. And it has been used by some parents who are dwarfs to ensure that their children will also be dwarfs, or by some parents who are deaf to ensure that their children are also deaf. There is no law against genetic screening of embryos for other traits, and it would be hard to imagine a law against a woman adjusting hormone levels during a pregnancy.

I can say what interventions I would find reasonable, but I am not the one to decide. That is for society at large. What I and other scientists can decide is whether or not to pursue knowledge that has the potential to teach us a great deal and to provide powerful tools that could be used for good purposes and bad. I am a hard-liner when it comes to knowledge. My work is all about finding things out, about learning, and I operate on the principle that we should try to find out as much as we can about the way the world works. I don't stop and say, Could this research find out something that might be misused, might cause more evil than good? I follow my nose to see what is interesting. When it comes to the question of how that knowledge is used, I am just another citizen.

Other hands will be trying to grow a dinosaur. But it's a project that I intend to support and campaign for in any way I can. If this book provokes a discussion, disagreement, and serious consideration about the project, great. If it helps get the research going, even better.

There is an image that keeps popping into my mind. I give an awful lot of lectures. I don't read from notes, I prefer to use slides, each of which fits with a topic that I want to talk about. I don't need to memorize a speech, or make it formal, I can stay conversational, which is what I find most comfortable.

So the image I have is that I walk onstage with a dinosaur on a leash. It's small, but bigger than a chicken. Let's say it's the size of turkey, one day maybe even the size of an emu. The dinosaur, or chickenosaur, or dinochicken, the emu-sized version of a dinosaur (that one might have a muzzle or a couple of handlers) is the ultimate slide. Instead of a lecture, this would be a public science class with questions and challenges about how it was done, what its skin feels like, does it have teeth, what does it eat, how close is it really to a dinosaur? What would inevitably follow would be a discussion about the nature of dinosaurs, of birds, of evolution and development, of the relationship of molecular biology to big changes in evolution, of how we know what we know, and whether we were justified in doing what we did. It would be something like this book, in conversational form.

That would be the most satisfying lecture I could possible give. I don't like providing answers. I never have. I like questions. I like asking them, trying to figure out answers, trying to figure out what we are really asking, and what new questions come up. For this event I won't have to prepare any speech at all. My entire prepared text will consist of one simple question, from which everything else will follow.

I'll walk to the edge of the stage, point to the creature on the leash, look at the audience, and say, "Can anyone here tell me what this is?"

APPENDIX:
CHICKENOSAURUS SKELETON

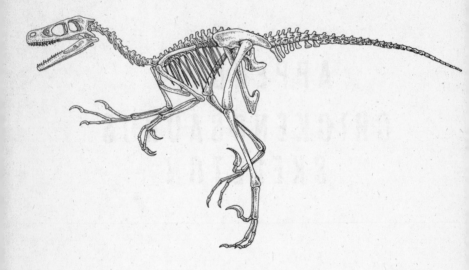

A quick look at the skeletons on these two pages would suggest that they are similar animals. No scientific expertise is necessary to see the two legs, the long neck and tail, the rib cage of each animal. Well, they are the same sort of animal; they are both dinosaurs. At left is *Saurornitholestes*, a small, theropod from the Cretaceous of North America. At right is *Chickenosaurus*, a small, so-far imaginary dinosaur from the future.

Birds, as I have been saying throughout the book, are dinosaurs, and these two drawings make it clearer than any words. *Saurornitholestes* is drawn as it was, or as best we know it, from fossils. *Chickenosaurus* is a rough sketch of the creature I am certain we can grow in the near future from a chicken egg. It was drawn by adding a tail and dinosaurlike arms to the skeleton of a chicken.

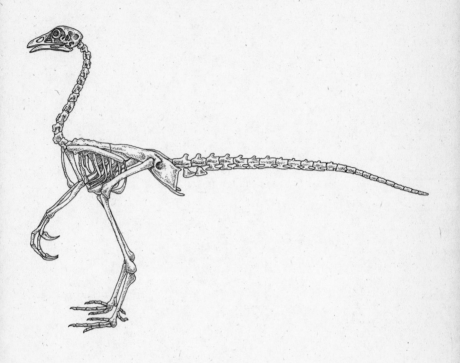

The arms on *Chickenosaurus* are clearly different from chicken wings and would require getting the forelimbs of an embryo to grow in a very different fashion. And there is the tail, which would be achieved by interfering with the instructions that stop backbone growth.

These changes would not make *Chickenosaurus* the same as *Saurornitholestes,* of course. There are differences in hip and chest structure, and in the skull, among others. But the skeletons show clearly the similarities and the small distance between the two creatures, which can be obscured by the familiar appearance of a living chicken and the way we think of dinosaurs.

BIBLIOGRAPHY

CHAPTER ONE

"Montana Freemen Disrupt Their Trial as Jury Is Picked." *The New York Times*, 1998, 12.

"*Tyrannosaurus rex:* The First Specimens of *Tyrannosaurus rex* Collected by the American Museum of Natural History with Previously Unpublished Information About Their Discovery."

"U.S. Census, 2000 Census, Population by Poverty Status in 1999 for Counties: 2000. Source: Census 2000 Sample Demographic Profiles, Table Dp-3."

Brooke, James. "Anti-Government Freemen Are Found Guilty of Fraud." *The New York Times*, 1998, 12.

———. "Behind the Siege in Montana, Bitter Trail of Broken Bonds." *The New York Times*, 1996, 1.

Brown, Barnum. "Expedition for Dinosaurs in Central Montana, 1908." Annual Reports of Paleontological Expeditions, The American Museum of Natural History (1908).

———. "Field Book, Barnum Brown, 1908, Hell Creek Beds, Montana." (1908).

DeVoto, Bernard, ed. *The Journals of Lewis and Clark*. Boston: Houghton Mifflin Company, 1997.

Dingus, Lowell. *Hell Creek, Montana: America's Key to the Prehistoric Past.* New York: St. Martin's Press, 2004, p. 242.

Egan, Timothy. "Siege Is Subplot in Town's Survival Drama." *The New York Times,* 1996, 1.

Fastovsky, David E., and David B. Weishampel. *The Evolution and Extinction of the Dinosaurs.* Cambridge, New York: Cambridge University Press, 1996.

Flannery, Tim. *The Eternal Frontier: An Ecological History of North America and Its Peoples.* New York: Grove Press, 2001.

Fowler, Loretta. *The Columbia Guide to American Indians of the Great Plains.* New York: Columbia University Press, 2003.

Goodstein, Laurie. "'Freemens'' Theological Agenda." *The Washington Post,* 1996.

Gould, Stephen J. *Wonderful Life: The Burgess Shale and the Nature of History.* New York: W. W. Norton, 1989.

Haines, Francis. *The Buffalo.* New York: Thomas Y. Crowell Company, 1970.

Hartman, Joseph H., Kirk R. Johnson, and Douglas J. Nichols, eds. "The Hell Creek Formation and the Cretaceous-Tertiary Boundary in the Northern Great Plains." Vol. Special Paper 361. Boulder, CO: The Geological Society of America, 2002.

Horner, Jack. *Dinosaurs Under the Big Sky.* Missoula, MT: Mountain Press, 2001.

Jakes, Dale and Connie. *The Buffalo.* Los Angeles: Dove Books, 1998.

Jordan, Arthur J. *Jordan.* Missoula, MT: Mountain Press, 2003.

Josephy, Alvin M., Jr. *The Indian Heritage of America.* Boston: Houghton Mifflin, 1991, p. 416.

Kemmick, Ed. "From Hell to Healing: Meth's Burden on Law Enforcement." *Billings Gazette,* 2005.

———. "Meth Summit Listens to Personal Tales of Horror." *Billings Gazette,* 2005.

King, Philip B. *The Evolution of North America.* Princeton, NJ: Princeton University Press, 1977.

Lott, Dale F. *American Bison: A Natural History.* Berkeley: University of California Press, 2002.

Malone, Michael P., Richard B. Roeder, and William L. Lang. *Montana: A History of Two Centuries*. Seattle and London: University of Washington Press, 1976.

Mayor, Adrienne. *Fossil Legends of the First Americans*. Princeton, NJ: Princeton University Press, 2005.

McHugh, Tom. *The Time of the Buffalo*. Lincoln: University of Nebraska Press, 1979.

Niebuhr, Gustav. "Creed of Hate Called Christian Identity Is the Heart of the Freemen's Beliefs." *The New York Times*, 1996, 14.

Prothero, Donald R. *After the Dinosaurs: The Age of Mammals*. Bloomington: Indiana University Press, 2006.

Robbins, Jim. "Siege Is Now Part of the Landscape." *The New York Times*, 1996, 14.

Schumacher, Otto L., and Lee A. Woodward. *Magnificent Journey: A Geologic River Trip with Lewis and Clark Through the Upper Missouri River Breaks National Monument*. Spokane, WA: Woodhawk Press, 2004.

Schweitzer, Mary H. "Notes from Montana, M. Schweitzer's Field Notes."

Toole, K. Ross. *Montana: An Uncommon Land*. Norman and London: University of Oklahoma Press, 1959.

Weil, Anne. "The Hell Creek Formation and the Cretaceous-Tertiary Boundary in the Northern Great Plains: An Integrated Continental Record of the End of the Cretaceous." *Journal of Paleontology* (2004): 1028–29.

CHAPTERS TWO AND THREE

Abelson, Philip H. "Paleobiochemistry." *Scientific American* 195 (1956): 83–92.

Asara, John M. "Response to Comment on 'Protein Sequences from Mastodon and *Tyrannosaurus rex* Revealed by Mass Spectrometry.'" *Science* 319 (2008): 33d.

Asara, John M., Mary H. Schweitzer, Lisa M. Freimark, Matthew Phillips, Lewis C. Cantley. "Protein Sequences from Mastodon and *Tyrannosaurus rex* Revealed by Mass Spectrometry." *Science* 316 (2007): 280–85.

Briggs, Derek E. G., and Amanda J. Kear. "Fossilization of Soft Tissue in the Laboratory." *Science* 259 (1993): 1439–42.

Buckley, Mike, et al. "Comment on 'Protein Sequences from Mastodon and *Tyrannosaurus rex* Revealed by Mass Spectrometry.'" *Science* 319 (2008): 33C.

Buehler, Markus J. "Nature Designs Tough Collagen: Explaining the Nanostructure of Collagen Fibrils." *Proceedings of the National Academy of Sciences* 103 (2006): 12283–90.

Chin, Karen, David A. Eberth, Mary H. Schweitzer, Thomas A. Rando, Wendy J. Sloboda, and John R. Horner. "Remarkable Preservation of Undigested Muscle Tissue Within a Late Cretaceous Tyrannosaurid Coprolite from Alberta, Canada." *Palaios* 18 (2003): 286–94.

Collins, M. J., C. M. Nielsen-Marsh, J. Hiller, C. I. Smith, and J. P. Roberts. "The Survival of Organic Matter in Bone: A Review." *Archaeometry* 44 (2002): 383–94.

Dal Sasso, Cristiano, and Marco Signore. "Exceptional Soft-Tissue Preservation in a Theropod Dinosaur from Italy." *Nature* 392 (1998): 383–87.

de Jong, E. W., P. Westbroek, J. F. Westbroek, and J. W. Bruning. "Preservation of Antigenic Properties of Macromolecules over 70 Myr." *Nature* 252 (1974): 63–64.

Gibbons, Anne. "Possible Dino DNA Find Is Greeted with Skepticism." *Science* 266 (1994): 1159.

Hedges, S. Blair, and Mary H. Schweitzer. "Detecting Dinosaur DNA: Technical Comments." *Science* 268 (1995): 1191.

Henikoff, Steven. "Detecting Dinosaur DNA." *Science* 268 (1995): 1192.

Hurd, Gary S. "Ancient Molecules and Modern Myths." The TalkOrigins Archive. http://www.talkorigins.org/faqs/dinosaur/osteocalcin.html (2004).

Knoll, Andrew H. "A New Molecular Window on Early Life." *Science* 285 (1999): 1025–26.

Laard, Marc W., Deshea Young, and Yentram Huyen. "Detecting Dinosaur DNA: Technical Comments." *Science* 268 (1995): 1192.

Nielsen-Marsh, Christina M., Michael P. Richards, Peter V. Hauschka, et al. "Osteocalcin Protein Sequences of Neanderthals and Modern Primates." *Proceedings of the National Academy of Sciences* 102 (2005): 44098–413.

Nielsen-Marsh, Christina M., Peggy H. Ostrom, Hasand Gandhi, et al. "Sequence Preservation of Osteocalcin Protein and Mitochondrial

DNA in Bison Bones Older Than 55 Ka." *Geology* 30, no. 12 (2002): 1099–102.

Ostrom, Peggy H., Hasand Gandhi, John R. Strahler, Angela K. Walker, Philip C. Andrews, Joseph Leykam, Thomas W. Stafford, Robert L. Kelly, Danny N. Walker, Mike Buckley, and James Humpula. "Unraveling the Sequence and Structure of the Protein Osteocalcin from a 42 Ka Fossil Horse." *Geochimica et Cosmochimica Acta* 70 (2006): 2034–44.

Peterson, Kevin J., Roger E. Summons, and C. J. Donoghue. "Molecular Palaeobiology." *Palaeontology* 50, Part 4 (2007): 775–809.

Poinar, Hendrik N., and Artur B. Stankiewicz. "Protein Preservation and DNA Retrieval from Ancient Tissues." *Proceedings of the National Academy of Sciences* 96 (1999): 8426–31.

Rowe, Timothy, and Earle F. McBride. "Dinosaur with a Heart of Stone: Technical Comment." *Science* 291 (2001): 783a.

Runnegar, Bruce. "Molecular Palaeontology." *Palaeontology* 29, Part I (1986): 1–24.

Russell, Dale A., Paul E. Fisher, Reese E. Barrick, and Michael K. Stoskopf. "Dinosaur with a Heart of Stone: Technical Comment, Response." *Science* 291 (2001): 783a.

Schweitzer, M. H., L. Chiappe, A. C. Garrido, J. M. Lowenstein, and S. H. Pincus. "Molecular Preservation in Late Cretaceous Sauropod Dinosaur Eggshells." *Proceedings of the Royal Society* B 2d72 (2005): 775–84.

Schweitzer, M. H., J. L. Wittmeyer, and J. R. Horner. "Gender-Specific Reproductive Tissue in Ratites and *Tyrannosaurus rex*." *Science* 308, no. 5727 (2005): 1456–60.

Schweitzer, M. H., J. L. Wittmeyer, J. R. Horner, and J. K. Toporski. "Soft-Tissue Vessels and Cellular Preservation in *Tyrannosaurus rex*." *Science* 307, no. 5717 (2005): 1952–55.

Schweitzer, Mary H., Jennifer L. Wittmeyer, John R. Horner, and Jan K. Toporski. "Soft-Tissue Vessels and Cellular Preservation in *Tyrannosaurus rex*." *Science* 307 (2005): 1952–55.

Schweitzer, Mary H., Mark Marshall, Keith Carron, D. Scott Bohle, Scott C. Busse, Ernst V. Arnold, Darlene Barnard, J. R. Horner, and Jean R. Starkey. "Heme Compounds in Dinosaur Trabecular Bone." *Proceedings of the National Academy of Sciences* 94 (1997): 6291–96.

Schweitzer, Mary H., Zhiyong Suo, Recep Avci, John M. Asara,

Mark A. Fernando, Teran Arce, and John R. Horner. "Analyses of Soft Tissue from *Tyrannosaurus rex* Suggest the Presence of Protein." *Science* 316 (2007): 277–80.

Schweitzer, Mary H., Jennifer L. Wittmeyer, and John R. Horner. "Gender-Specific Reproductive Tissue in Ratites and *Tyrannosaurus rex*." *Science* 208 (2005): 1456–60.

Schweitzer, Mary Higby. "The Future of Molecular Paleontology." *Palaeontologica Electronica* 5 (2003) Web site, http://palaeo-electronica.org.

Schweitzer, Mary Higby, Christopher L. Hill, John M. Asara, William S. Lane, and Seth H. Pincus. "Identification of Immunoreactive Material in Mammoth Fossils." *Journal of Molecular Evolution* 55 (2002): 696–705.

Torres, Jesus M., Concepción Borja, and Enrique G. Olivares. "Immunoglobulin G in 1.6 Million-Year-Old Fossil Bones from Venta Micena (Granada, Spain)." *Journal of Archaeological Science* 29 (2002): 1–75.

Weiner, S., H. A. Lowenstam, and L. Hood. "Characterization of 80-Million-Year-Old Mollusk Shell Proteins." *Proceedings of the National Academy of Sciences* 73 (1976): 2541–45.

Woodward, Scott R. "Detecting Dinosaur DNA: Technical Comments, Response." *Science* 268 (1995): 1194.

Woodward, Scott R., Nathan J. Weyand, Bunnell. "DNA Sequence from Cretaceous Period Bone Fragments." *Science* 266 (1994): 1229.

Wyckoff, Ralph W. G., Estelle Wagner, Philip Matter III, and Alexander R. Doberenz. "Collagen in Fossil Bone." *Proceedings of the National Academy of Sciences* 50 (1963): 215–18.

Yeoman, Barry. "Schweitzer's Dangerous Discovery." *Discover* magazine, 2006.

Zischler, H., M. Hoss, O. Handt, A. von Haeseler, A. C. van der Kuyl, J. Goudsmit, and S. Paabo. "Detecting Dinosaur DNA: Technical Comments." *Science* 268 (1995): 1192–93.

CHAPTER FOUR

"Are Birds Really Dinosaurs?" DinoBuzz, http://www.ucmp.berkeley.edu/diapsids/avians.html.

Chen, Pei-Ji, Zhi-ming Dong, and Shuo-nan Zhen. "An Exceptionally Well-

Preserved Theropod Dinosaur from the Yixian Formation of China." *Nature* 391 (1998): 147–52.

Chiappe, Luis M. *Glorified Dinosaurs: The Origin and Early Evolution of Birds.* New York: John Wiley, 2007.

Norell, Mark, Quiang Ji, Kequin Gao, Chongxi Yuan, Yiin Zhao, and Lixia Wang. "'Modern' Feathers on a Nonavian Dinosaur." *Nature* 416 (2002): 36.

Ostrom, John. "The Ancestry of Birds." *Nature* 242 (1973): 136.

———. "The Origin of Birds." *Annual Review of Earth and Planetary Science* 3 (1975): 55–77.

Padian, Kevin. "When Is a Bird Not a Bird?" *Nature* 393 (1998): 729–30.

Pennisi, E. "Bird Wings Really Are Like Dinosaurs' Hands." *Science* 307 (2005): 194–95.

Prum, Richard O., and Alan H. Brush. "Developmental and Evolutionary Origin of Feathers." *Journal of Experimental Zoology* 285 (1999): 291–306.

———. "Which Came First, the Feather or the Bird?" *Scientific American* (2003): 86–93.

Qiang, Ji, Philip J. Currie, Mark A. Norell, and Ji Shu-an. "Two Feathered Dinosaurs from Northeastern China." *Nature* 393 (1998): 753–61.

Unwin, D. M. "Feathers, Filaments, and Theropod Dinosaurs." *Nature Reviews/Genetics* 391 (1998): 119–21.

Williston, S. W. "Birds." *The University Geological Survey of Kansas* 4, Part II (1898): 43–53.

Xu, Xing, Xiao-Lin Wang, and Xiao-Chun Wu. "A Dromaeosaurid Dinosaur with a Filamentous Integument from the Yixian Formation of China." *Nature* 401 (1999): 262–66.

Xu, Xing, Zhi-lu Tang, and Xiao-lin Wang. "A Therizinosauroid Dinosaur with Integumentary Structures from China." *Nature* 399 (1999): 350–54.

Xu, Xing, Zhonge Zhou, Xiaolin Wang, Xuewen Kuang, Fucheng Zhang, and Xiangke Du. "Four-Winged Dinosaurs from China." *Nature* 42 (2003): 336–40.

Zhao, Xijin, and Xing Xu. "The Oldest Coelurosaurian." *Nature* 394 (1998): 234–35.

CHAPTER FIVE

Carroll, Sean B. "Chance and Necessity: The Evolution of Morphological Complexity and Diversity." *Nature Reviews/Genetics* 409 (2001): 1102–109.

———. "The Big Picture." *Nature Reviews/Genetics* 409 (2001): 669.

———. *Endless Forms Most Beautiful: The New Science of Evo Devo*. New York: W. W. Norton, 2006.

———. *The Making of the Fittest: DNA and the Ultimate Forensic Record of Evolution*. New York: W. W. Norton, 2007.

Carroll, Sean, Jennifer Grenier, and Scott Weatherbee. *From DNA to Diversity: Molecular Genetics and the Evolution of Animal Design*. Malden, MA: Wiley-Blackwell, 2004.

Chuong, Cheng-Ming, Rajas Chodankar, Randall B. Widelitz, and Ting-Xin Jiang. "Evo-Devo of Feathers and Scales: Building Complex Epithelial Appendages." *Current Opinion in Genetics and Development* 10 (2000): 449–56.

Davidson, Eric H., and Douglas H. Erwin. "Gene Regulatory Networks and the Evolution of Animal Body Plans." *Science* 311 (2006): 796–800.

Gilbert, Scott F. DevBio: A Companion to *Developmental Biology, Eighth Edition*. "16.5 Did Birds Evolve from Dinosaurs?" (2003) Web site, http://www. Devbio.Com.

———. *Developmental Biology*. 7th ed. Sunderland, MA: Sinauer Associates, 2003.

Gould, Stephen J. *Ontogeny and Phylogeny*. Cambridge, MA: Harvard University Press, 1977.

Harré, Rom. *Great Scientific Experiments*. Oxford: Phaedon Press Limited, 1981.

Harris, Matthew P., Scott Williamson, John F. Fallon, Hans Meinhardt, Richard O. Prum. "Molecular Evidence for an Activator-Inhibitor Mechanism in Development of Embryonic Feather Branching." *Proceedings of the National Academy of Sciences* 102, no. 33 (2005): 11734–39.

International Chicken Genome Sequencing Consortium. "Sequence and Comparative Analysis of the Chicken Genome Provide Unique Perspectives on Vertebrate Evolution." *Nature* 432 (2004): 695–716.

Larsson, Hans C. E. "Hatching of an Accountable Science." *Science* 307 (2005): 520.

Lawrence, Peter A. *The Making of a Fly: The Genetics of Animal Design*. Oxford and Boston: Blackwell Scientific Publications, 1992.

Nikbaht, Neda, and John C. McLachlan. "Restoring Avian Wing Digits." *Proceedings of the Royal Society* B 266 (1999): 1101–104.

Prum, Richard O., and Alan H. Brush. "Developmental and Evolutionary Origin of Feathers." *Journal of Experimental Zoology* 285 (1999): 291–306.

———. "Which Came First, the Feather or the Bird?" *Scientific American* (2003): 86–93.

Raff, Rudolf A. "Evo-Devo: the Evolution of a New Discipline." *Nature Reviews/Genetics* 1 (2000): 74–79.

Wade, Nicholas. "The Fly People Make History on the Frontiers of Genetics." *The New York Times*, 2000, 1.

Wade, Nicholas. "Dainty Worm Tells Secrets of the Human Genetic Code." *The New York Times*, 1997, 1.

Wagner, Günter P. "What is the Promise of Developmental Evolution? Part II: A Causal Explanation of Evolutionary Innovations May be Impossible." *Journal of Experimental Zoology* 291 (2001): 305–309.

———. "The Developmental Evolution of Avian Digit Homology: An Update." *Theory in Biosciences* 124 (2005): 165–83.

Wagner, Günter P., and Jacques A. Gauthier. "1,2,3=2,3,4: A Solution to the Problem of the Homology of the Digits in the Avian Hand." *Proceedings of the National Academy of Sciences* 96 (1999): 5111–16.

Wagner, Günter P., and Hans C. E. Larsson. "What Is the Promise of Developmental Evolution? III. The Crucible of Developmental Evolution." *Journal of Experimental Zoology* 300B (2003): 1–4.

Wolpert, Lewis. *The Triumph of the Embryo*. Oxford: Oxford University Press, 1993.

Zimmer, Carl. *At the Water's Edge: Fish with Fingers, Whales with Legs, and How Life Came Ashore but Then Went Back to Sea*. New York: Touchstone/Simon & Schuster, 1999.

CHAPTER SIX

Dubrow, Terry J., Phillip Ashley Wackym, and M. A. Lesavoy "Detailing the Human Tail." *Annals of Plastic Surgery* 20, no. 4 (1988): 340–44.

Larsson, Hans C. E., and Günter P. Wagner. "Pentadactyl Ground State of the Avian Wing." *Journal of Experimental Zoology* (Mol Dev Evol) 294 (2002): 146–51.

Nikbaht, Neda, and John C. McLachlan. "Restoring Avian Wing Digits." *Proceedings of the Royal Society* B 266 (1999): 1101–04.

CHAPTER SEVEN

Resko, John A., Anne Perkins, Charles E. Roselli, James A. Fitzgerald, Jerome V. A. Choate, Fredrick Stormshak. "Endocrine Correlates of Partner Preference Behavior in Rams." *Biology of Reproduction* 55 (1996): 120–26.

ACKNOWLEDGMENTS

This book would not have been possible without the hard work and dedication of a great many people, foremost among them my former graduate student Dr. Mary Schweitzer, my colleague Dr. Hans Larsson, and former graduate student Chris Organ. A special thanks to each. Other paleontology colleagues whom I've been doing a lot of work with these past few years include Mark Goodwin and Kevin Padian, both of Berkeley, and I owe a debt of gratitude to each. I also thank my staff, including Pat Leiggi, Bob Harmon, Carrie Ancell, Ellen Lamm, Jamie Jette, and Linda Roberts; my former and current graduate students; and the many volunteers who gave up their summers to help excavate dinosaur specimens such as "B. rex." Thanks to Nathan Myhrvold, who provided the major funding for the excavation of B. rex, and some of the funding for our new mobile molecular field station. I also thank all my other funders, especially Tom and Stacey Seibel, Klein and Karen Gilhousen, Gerry Ohrstrom, Bea Taylor, the Paul Prager family,

Catherine Reynolds, George Lucas, and the Sands families. A very special thanks to Jim Gorman who so eloquently transformed my scientific lingo into literature. Finally, I want to thank my friend Carlye Cook for reminding me that science, especially of the sort discussed in this book, has an obligation to explain itself, in layman's terms, to the general public.

Both Jim and I also want to thank our agent, Kris Dahl, and our editor, Stephen Morrow, for their great work on the book; Oceana Gottlieb for a fantastic book jacket; Craig Schneider, the copy editor; and Julia Gilroy, production editor, for their care and attention in moving the manuscript to book form.

—Jack Horner

First and foremost, I want to thank my wife, Kate, who had the most difficult job I can imagine: living in the same house with me as I tried to steal every available minute and some that were not available to work on the book. Her support was invaluable, as was my son Daniel's patience with his father's day job and night and weekend job. My daughters, Madeleine and Celia, did not have to live in the same house, but they encouraged me from a distance. And my parents, as always, cheered me on.

I'm also grateful to friends and coworkers who encouraged me to push on and finish the book, or at least stop complaining about it.

And finally, to Jack, for all your intriguing research, for your open and inventive mind, for thinking of this book and of me to write it, thank you.

—Jim Gorman

INDEX

Note: Page numbers in *italics* indicate photographs and illustrations.